ELECTRONICS SWITCHING

ELECTRONICS SWITCHING

By

ANINDITA SAINI

Department of Electronics and Communication Engg.
S.D. College of Engineering & Technology,
Muzzafarnagar
Uttar Pradesh

UNIVERSITY SCIENCE PRESS

(An Imprint of Laxmi Publications Pvt. Ltd.)

An ISO 9001:2015 Company

BENGALURU • CHENNAI • GUWAHATI • HYDERABAD
KOCHI • KOLKATA • LUCKNOW • RANCHI
NEW DELHI

Printed and bound in India

First Edition : 2025

ISBN : 978-81-31807-97-2

Branches

✆	Bengaluru	080-26 75 69 30
✆	Chennai	044-24 34 47 26
✆	Guwahati	0361-254 36 69
✆	Hyderabad	040-27 55 53 83
✆	Kochi	0484-405 13 03
✆	Kolkata	033-40 04 77 79
✆	Lucknow	0522-430 36 13
✆	Ranchi	0651-224 24 64

PUBLISHED IN INDIA BY

UNIVERSITY SCIENCE PRESS

(An Imprint of Laxmi Publications Pvt. Ltd.)

An ISO 9001:2015 Company

113, GOLDEN HOUSE, GURUDWARA ROAD, DARYAGANJ,

NEW DELHI - 110002, INDIA

Corporate Office : B-33, Sector-5, Noida,

Gautambuddha Nagar (U.P.) - 201 301, INDIA

Telephone : +91-120-44 46 900

www.laxmipublications.com

info@laxmipublications.com

CONTENTS

CHAPTER 1.	**INTRODUCTION**	....	**1–14**
1.1	Introduction	....	1
1.2	Evolution of Switching System	....	1
1.3	Classification of Switching System	....	2
1.4	Message Switching	....	2
1.5	Circuits Switching	....	4
1.6	Comparison of Message Switching and Circuit Switching	....	6
1.7	Functions of a Switching System	....	6
1.8	Register-translator Sender	....	7
1.9	Distribution Frames	....	8
1.10	Crossbar Switch	....	9
1.11	A General Trunking	....	10
1.12	Electronic Switching	....	11
1.13	Reed Relay Electronic System	....	11
1.14	Digital Switching System	....	12
	Summary	....	13
	Review Questions	....	14
	Multiple Choice Questions	....	14
CHAPTER 2.	**DIGITAL SWITCHING**	....	**15–48**
2.1	Introduction	....	15
2.2	Switching Functions	....	15
2.3	Space Division Switching	....	16
2.4	Multiple Stage Switching	....	16
2.5	Non-blocking Switches	....	22
2.6	Blocking Probabilities	....	22
2.7	Comparison of Single Stage and Multistage Networks	....	24
2.8	Folded Four-Wire Switches	....	24
2.9	Path Finding	....	25
2.10	Switch Matrix Control	....	25
2.11	Time Division Switching	....	27
2.12	Analog Time Division Switching	....	27
2.13	Digital Time Division Switching	....	29
2.14	A Digital Memory Switch	....	31
2.15	Time Stages in General	....	32
2.16	Two Dimensional Switching	....	32

2.17 Implementation Complexity of TD Switches 33
2.18 Multiple Stage Time and Space Switching 34
2.19 No. 4 ESS Toll Switch 38
2.20 System 75 Digital PBX 39
2.21 Digital Cross Connect System 40
2.22 Consolidation and Segregation 41
2.23 DCS Hierarchy 42
2.24 Integrated Cross Connect Equipment 42
2.25 Digital Switching in an Analog Environment 43
Summary 46
Review Questions 47
Multiple Choice Questions 47

CHAPTER 3. TELECOM TRAFFIC ENGINEERING 49–71
3.1 Introduction 49
3.2 Network Traffic Load and Parameters 49
3.3 Grade of Service and Blocking Probability 53
3.4 Modelling Switching Systems 54
3.5 Markov Processes 56
3.6 Birth-death Processes 57
3.7 Incoming Traffic and Service Time Characterisation 59
3.8 Poisson Arrival Process 60
3.9 Holding Time of Calls 60
3.10 Blocking Models and Loss Estimates 61
3.11 Delay System 66
Summary 70
Review Questions 70
Multiple Choice Questions 70

CHAPTER 4. CONTROL OF SWITCHING SYSTEMS 72–92
4.1 Introduction 72
4.2 Call Processing Functions 72
4.3 Sequence of Operations 73
4.4 Signal Exchanges 73
4.5 State-Transition Diagram 74
4.6 Common Control 75
4.7 Reliability, Security and Availability 76
4.8 Stored Program Control (SPC) 76
4.9 Centralised SPC 77
4.10 Comparison of Three Modes of Centralised SPC 81
4.11 Distributed SPC 81

	4.12 SPC Software	....	85
	4.13 System Software and Language Processor	....	85
	4.14 SDL (Specification Description Language)	....	87
	4.15 Application Software	....	89
	Summary	....	91
	Review Questions	....	92
	Multiple Choice Questions	....	92
CHAPTER 5.	**SIGNALLING**	....	**93–115**
	5.1 Introduction	....	93
	5.2 Customer Line Signalling	....	93
	5.3 AF (Audio Frequency) Junctions and Trunk Circuits	....	95
	5.4 Classification of Signalling	....	97
	5.5 Inchannel Signalling	....	97
	5.6 PCM Signalling	....	100
	5.7 Inter-register Signalling	....	101
	5.8 Common Channel Signalling (CCS)	....	103
	5.9 Difference Between Inchannel and Common Channel Signalling	....	107
	5.10 CCITT Signalling System No. 6 (SS6)	....	107
	5.11 CCITT Signalling System No. 7 (SS7)	....	108
	5.12 HDLC Protocol	....	111
	5.13 Signal Units	....	111
	5.14 Signalling Information Field (SIF)	....	113
	Summary	....	113
	Review Questions	....	114
	Multiple Choice Questions	....	114
CHAPTER 6.	**PACKET SWITCHING**	....	**116–140**
	6.1 Introduction	....	116
	6.2 Packet Switching	....	116
	6.3 Statistical Multiplexing	....	118
	6.4 Routing Control	....	119
	6.5 Flow Control	....	123
	6.6 X.25 Protocol	....	123
	6.7 Frame Relay	....	124
	6.8 Difference Between Frame Relay and X.25 Packet Switching	....	126
	6.9 TCP/IP	....	126
	6.10 ATM (Asynchronous Transfer Mode)	....	132
	6.11 Advantages of ATM	....	138
	Summary	....	139
	Review Questions	....	139
	Multiple Choice Questions	....	140
	Index	....	**141**

Dedicated to

my

Parents

ACKNOWLEDGEMENT

I wish to express my profound thanks to the people around me whose help make this book a reality. But it would be unfair to forget GOD whose blessing forward me to write this book.

First of all I wish to express my sincere gratitude to Dr. S.N. Chauhan, Director of S. D. College of Engineering & Technology, Muzzafarnagar for providing encouraging environment, without which this book would not have come into existence.

I am extremely thankful to Dr. Arvind Kumar Gautam, Principal of S.D. College of Engineering & Technology, Muzzafarnagar and Prof. N.K. Sharma, HOD of Electronics & communication.

I wish to express my regards to Mrs. Pragati Sharma, Mrs. Paural Gupta, Mrs. Vaishali Khosla, Miss Prachi Agarwal, Miss Priyanka Saini, Mr. Jitendra Sharma, Mr. Debraj, Mr. S. K. Singh and all other colleagues.

I record with gratitude the moral support received from my family members specially from my husband Mr. Manish Saini.

I thankfully acknowledge the contribution of various authors of different books, and magazines from where materials have been collected to enrich the contents of the book.

—Author

PREFACE

It is a great pleasure to present this book to the students and readers. The book fully covers the syllabus of various Indian universities and the institution of Engineering. The whole volume has been divided into 6 chapters.

Chapter 1 & 2 deals with the computer controlled switching procedures and digital switching systems.

Chapter 3 discuss the teletraffic engineering system.

Chapter 4 devoted to SPC exchanges those are basics for digital switching system.

Chapter 5 deals with signalling techniques. The signalling system enables the quick path setup between calling & called subscribers.

Entire 6th chapter discusses the idea about ATM, OSI Model, TCP/IP concepts.

Many persons have contributed directly or indirectly to this book. All these are gratefully acknowledged.

I don't claim the work as perfect without any mistake or errors, therefore, I request the readers and teaching community to bring to the notice of publishers and author both any printing error which can be rectified in next editions.

—Author

KEY FETURE OF BOOK

- Thorough coverage of all the topics.
- More detail & modular representation of chapters.
- Multiple choice questions & model papers are also included except review questions.
- Simple & user friendly language.

1

Introduction

1.1 INTRODUCTION

In present scenario, universal connectivity like Internet, Telecommunication becomes a fabric of our life. Switching systems are designed for voice as well as data communication with the help of the telecommunication links and computers. The main purpose of the switching system is to provide the means to pass information from any terminal device to other terminal device selected by the originator. Switching systems come into existence as a result of rapid growth of electronic systems. In this chapter, we discuss some basics related to electronic switching system that is used to provide worldwide connectivity.

1.2 EVOLUTION OF SWITCHING SYSTEM

In early days, human operated manual switching system were used. After recognising the relative merits and demerits of a manual switching system, an automatic switching system has come into existence.

Automatic switching system can be classified as electromechanical and electronic switching system. The electromechanical switching further divided into step-by-step or stronger system and crossbar system. The step-by-step switching system invented by Almon. B. Strowger in 1892 and in 1938, AT and T laboratories in US introduced crossbar switching system in the field. Then comes to computer controlled switching system which replace the electromechanical switching system, referred to as Stored Program Control (SPC). The first computer controlled switching was introduced in 1960. Transistors and printed circuit board technology were used in computer controlled switching till 1965. After that microprocessor come into the field.

Computer controlled switching led to refer as Electronic Switching System (ESS) or Electronic Automatic Exchange (EAX). Digital electronic switching were first introduced in the US Public Network in 1976. With AT & T and by the mid 1980's electronic switching has changed to almost digital.

A simplified diagram of switching system is shown in Fig. 1.1, where subscribers are connected to each other through the switching system.

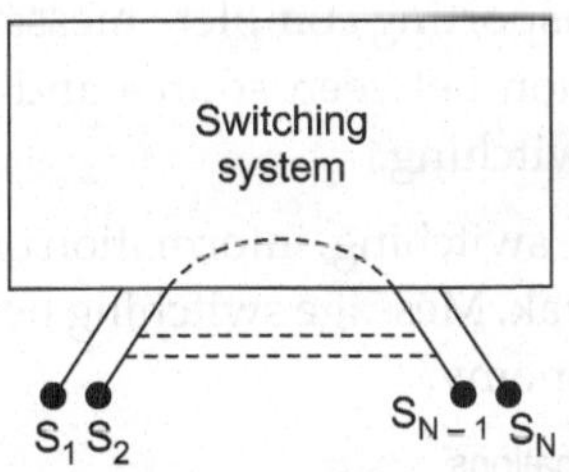

Fig. 1.1 *Subscribers interconnection using a switching system*

1.3 CLASSIFICATION OF SWITCHING SYSTEM

Classification of switching system is shown in Fig. 1.2 below :

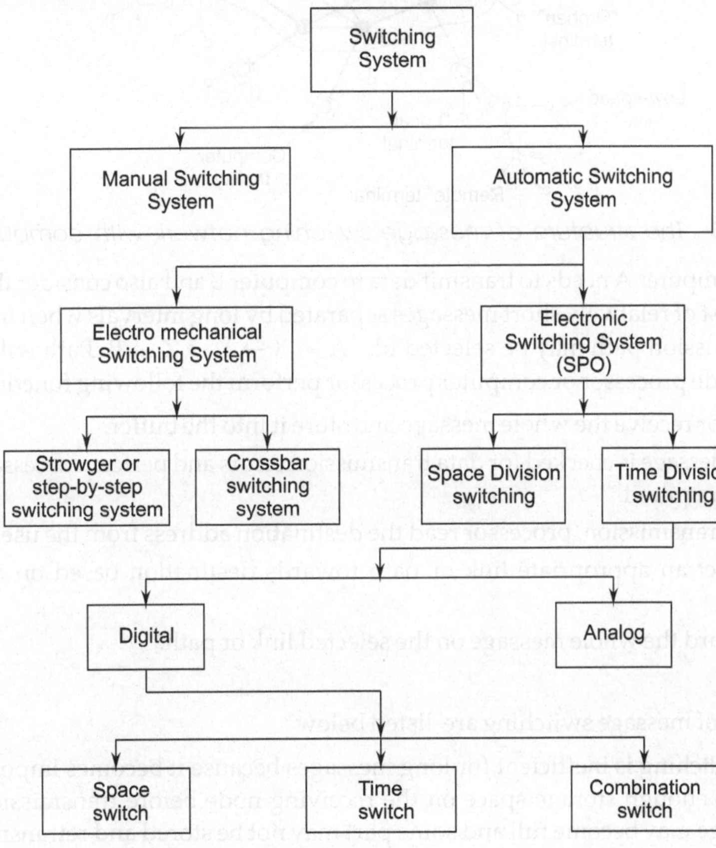

Fig. 1.2 *Classification of switching system*

1.4 MESSAGE SWITCHING

The process of transferring messages from one place or point to other place or point then it is known as switching. In case of telecommunication, message switching and circuit switching are used.

Another type of switching *i.e.,* packet switching is used in computer communication network. Message switching and circuit switching are discussed in this chapter but packet switching is explained in the Chapter No. 6. Transporting complete messages from a source to a destination in non-real time and without interaction between source and destination, usually in a store and forward fashion is called message switching.

In short we can say that in message switching, information or data or message is transmitted from source to destination without any break. Message switching network such as represented in Fig. 1.3, offers the possibility of improved economy.

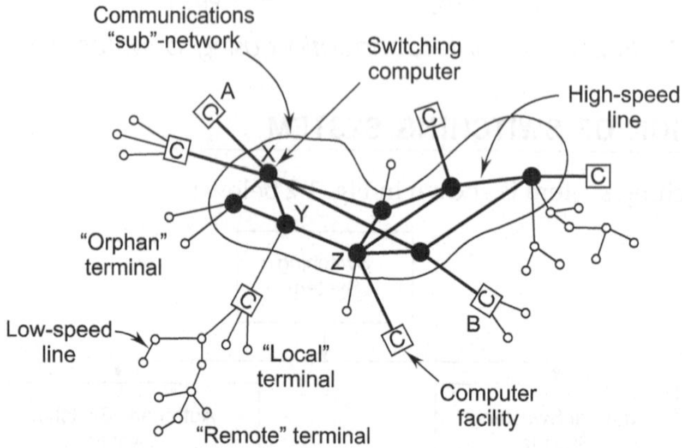

Fig. 1.3 *The structure of message switching network with computers*

Suppose that computer A needs to transmit data to computer B and also consider that the communication will consist of relatively short messages separated by long intervals when there is no transmission. A transmission path may be selected like A → X → Y → Z → B. Path will be as short as possible. Every node processor or computer processor perform the following functions :

- The processor receive the whole message and store it into the buffer.
- The whole message is checked for data transmission errors and perform necessary error recovery action if required.
- For further transmission, processor read the destination address from the user message.
- Then it select an appropriate link or path towards destination based on certain routing criterion.
- At last forward the whole message on the selected link or path.

Disadvantages

Disadvantages of message switching are listed below :

- Message switching is inefficient for long messages because it becomes important to ensure that there is enough storage space on the receiving node before transmission. Otherwise, buffer storage may become full and some part may not be stored and retransmission of message will be necessary.
- During transmission, if any error occurs, the whole message may have to be retransmitted.
- If a high priority short message arrives while a long message is in transmission, it will have to wait until the transmission of long message ends. Therefore, delay occurs.

- Retransmission of long message creates traffic in the network.

These drawbacks are overcome in case of Packet switching.

Note. Since, the system of message switching involves storing and subsequent forwarding of messages, a message switch network is referred to as a "store and forward network".

1.5 CIRCUITS SWITCHING

In case of circuits switching, a physical connection is established between the source and the destination without interruption for the entire duration of the transmission. This type of system is entirely analogous to a telephone switching system. The entire path remains dedicated to the single set of users until transmission and other necessary actions completed and during this time no any other user can use the same path even during intervals also when path remains idle. So, in order to set up a direct connection over many links or paths it is necessary that each path to be simultaneously free. The physical connection is done by simple wires or co-axial cables or radio or satellite links. Fig. 1.4 illustrates the major principle of circuit switching. Generally circuit switching is accomplished by TDM (Time division multiplexing).

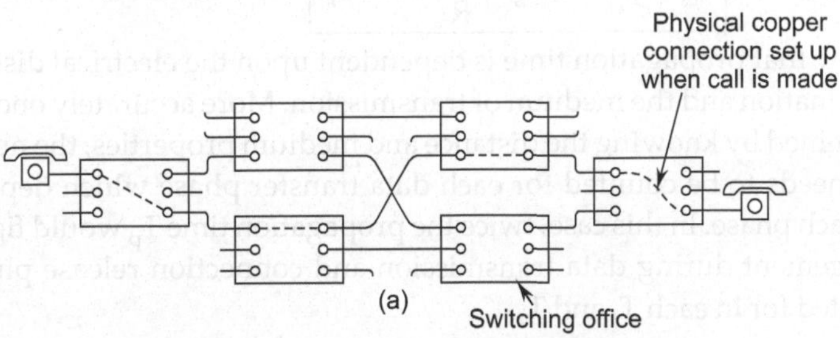

Fig. 1.4 *General Circuit Switching*

In the Fig. (1.4) a suitable path is established by physical wire connections and maintained until disconnected by the user.

In the circuit switching three major functions are performed as follows :

- The source request the network for the route before transmission. The network selects a route.
- Now the data transfer takes place. The duration of data transfer is called holding time.
- After completion of the data transfer, the path set up is disconnected and the path becomes free for another transmission.

So, total time taken by circuit switched connection is expressed as,

$$\boxed{T_{cs} = T_e + T_t + T_r} \quad ...(1.1)$$

where,

T_e = time for connection establish

T_t = time for data transmission

T_r = time for path release.

Now, T_e depends upon the number of switching nodes in the path between source and destination hosts. Each node except the last one to which the destination host is connected take sometime to select the path by routing algorithm. So T_e may be expressed as,

$$\boxed{T_e = (N-1)\, T_s} \quad ...(1.2)$$

where, N = number of switching nodes in the path.

T_s = average route selection time in each node.

T_t depends upon the data rate and the size of the message so T_t may be written as,

$$\boxed{T_t = \frac{M}{R}} \quad ...(1.3)$$

where, M = message length in bits

R = data rate in bps.

Again T_r depends upon the number of switching nodes in the path and sometime for making housekeeping entries for each node like making entries in routing tables after getting the release signal. So,

$$\boxed{T_r = N \cdot T_n} \quad ...(1.4)$$

where, N = number of switching node in the path

T_n = time for making housekeeping entries.

Therefore we have,

$$\boxed{T_{cs} = (N-1)\,T_s + \frac{M}{R} + N \cdot T_n} \quad ...(1.5)$$

Now, as we know that propagation time is dependent upon the electrical distance between the source and the destination and the medium of transmission. More accurately one way propagation time may be determined by knowing the distance and medium properties, the number of times the propagation time needs to be counted for each data transfer phase which depend on the actual protocols used in each phase. In this case, twice the propagation time T_p would figure in T_e. Assuming no acknowledgement during data transmission and connection release phases, only one T_p needs to be accounted for in each T_t and T_r.

Therefore equation (1.5) can be written as,

$$\boxed{T_{cs} = (N-1)\,T_s + \frac{M}{R} + N \cdot T_n + 4T_p} \quad ...(1.6)$$

where T_p is one way propagation time.

Example 1.1. *A circuit switched connection involves 4 switching nodes. Each node takes 1 seconds and 1 second for establishing and releasing connections respectively. If the data transfer rate is 2800 bps, compute the data transfer time for a message that is 500 bytes long.*

Solution. By neglecting propagation delay time, the total data transfer time will be,

$$T_{cs} = (N-1)\,T_s + \frac{M}{R} + N \cdot T_n \;[T_p = 0]$$

$$= (4-1)\,1 + \frac{500 \times 8}{2800} + 4 \times 1 = 7.7$$

Disadvantages

The disadvantages of circuits switching is listed below :

- The path set up time which is typically of the order of 20–30 seconds or more, turns out to be an excessive overhead for computer traffic which generally lasts for a few seconds or less.
- The entire line quality is affected if there is one bad link in the network as it will be referred poor quality although major portion of the network uses high quality links. This type of problem is termed as 'last mile' problem.

- The speed of operation of the circuit is low.
- The required bandwidth is statically allocated and the unused bandwidth is wasted.
- There is no error control facilities in the circuit. The errors are handled by the end system.

Note. Circuit switching is inefficient for small messages where the total data transfer time is dominated by the path setup time.

1.6 COMPARISON OF MESSAGE SWITCHING AND CIRCUIT SWITCHING

Message switching	*Circuit switching*
1. In message switching there is no need of establishing physical path.	1. In circuit switching, physical path establishment is the main requirement.
2. The source and destination never interact in real time.	2. The source and destination are connected temporarily during data transfer.
3. Suitable for short message transfer.	3. Suitable for long message transfer.
4. Transmission paths are never idle.	4. After path set up if the users denied service, the line will be idle.
5. Destination node status is not required before sending message.	5. Destination node status is important to know before set up the path for transmission.
6. Message switching network generally accepts all traffic but provides longer time delivery because of increased queue length.	6. Circuit switching network rejects over traffic if all the lines are busy.

1.7 FUNCTIONS OF A SWITCHING SYSTEM

The switching system or exchange, in case of telecommunication perform the main function in order to connect two subscribers. The system receives the control signals, messages or conversations and forwards to the required destination after modification if necessary. There basic functions are divided in some sub-functions as follows :

1. Attending
2. Information receiving
3. Information processing
4. Busy testing
5. Interconnection
6. Alerting
7. Supervision
8. Information sending.

1. **Attending :** The system must attend all the call requests and for that all the lines connected to the system must be monitored continuously. The calling signal also known as seize signal as it obtains a resource from the exchange or switching system.
2. **Information receiving :** After attending the call, the system receive the destination address or the required called line from the caller or calling subscriber.
3. **Information processing :** The system must process the received information for further actions to be performed and to control these actions. In addition to address information originating and terminating calls or service information also processed.
4. **Busy testing :** Before forwarding the information, the system must make a busy test to know whether the line is free or already engaged to another call.

5. **Interconnection :** Connections are made between two customers on call in the following sequence
   (*a*) A connection to the calling terminal
   (*b*) A connection to the called terminal
   (*c*) A connection between the two terminals.

   Some modern, automatic switching systems release the initial connections (*a*) and (*b*) and establish connection (*c*) over a separate path through the switching network. This is termed as call back or crank back.
6. **Alerting :** When connection is made, the system sends an alert signal to the called subscriber for the call *i.e.,* by sending ringing current to a called subscriber's telephone.
7. **Supervision :** During conversation, the switching system continuously monitor the connection in order to clear it down when the call has ended.
8. **Information sending :** If the called subscriber's line is located on another exchange or switching system the additional function of information must be sent. The originating exchange must send the address of the terminating exchange.

Note. In telephonic conversation, the one who initiates the call is known as calling subscriber and the one for whom the call is destined is called subscriber.

1.8 REGISTER-TRANSLATOR SENDER

This term is used in common control switching system (applicable to crossbar switching). The block diagram of a common control switching system is shown in Fig. 1.5.

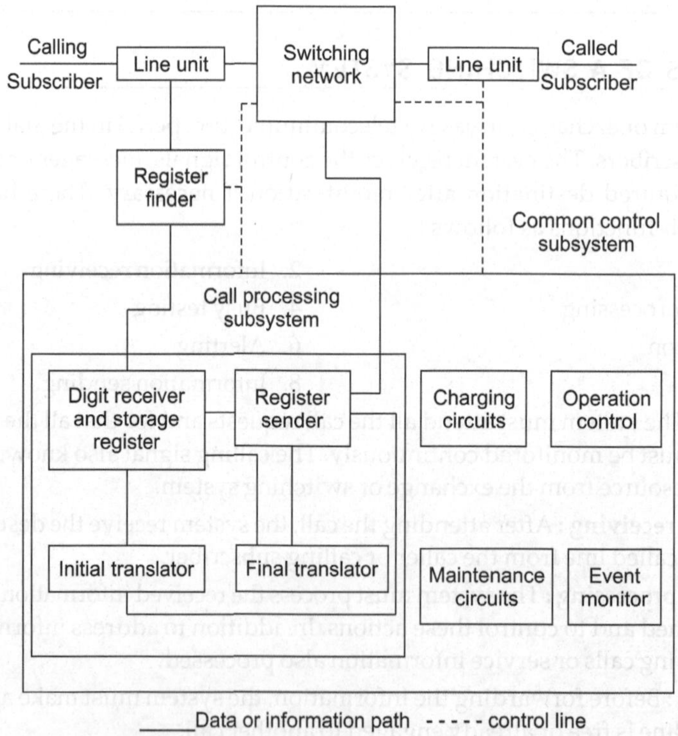

Fig. 1.5 *Common Control Switching System*

When a subscriber is in off hook condition, (*i.e.*, pick-up the handset) the switching system or the exchange gets the calling location and marked for dial tone. Register finder (shown in Fig.) is activated to seize the free register and identify the category of the dialling (*i.e.*, pulse dialling or multifrequency dialling). An appropriate free register according to the line which is based on the dialling is chosen and the register sends the dial tone to the calling subscriber. After that, when the subscriber dial the initial digits (generally 2 to 5) which identify the exchange goes to the register. As soon as possible the register sends this numbers to the initial translator for further processing. The remaining numbers goes to the register.

The initial translator select the route of the network and decides whether the call is possible to establish or not. It also determine the charging method. If there is any type of fault in the trunk lines then initial translator decides the other route for communication.

If the destination or called number belong to the other exchange then the initial translator generates the required routing digits and passes them to the register-sender. The register-sender send the number to the destination by using proper signalling technique, according to the requirement of the destination exchange. If the called number belongs to the same exchange or present exchange then the digits are processed by the final translator which decides the line and category of the called line.

1.9 DISTRIBUTION FRAMES

Every subscriber in a telephone network is connected generally to the nearest switching office by means of a dedicated pair of wires. The subscriber's lines, meters, answering and multiple jacks and other associated circuits are all interconnected via two distribution frames called the main distribution frame (MDF) and the intermediate distribution frame (IDF). The Fig. 1.6 illustrates the distribution frame.

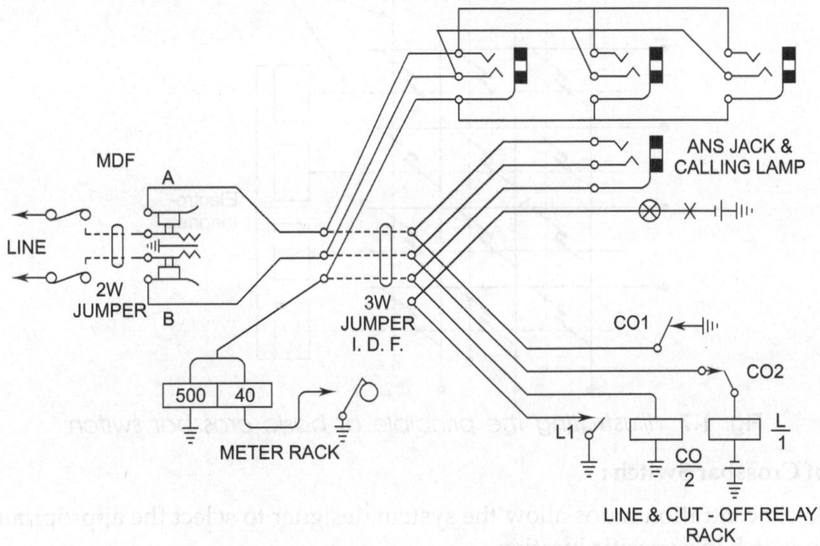

Fig. 1.6 *Main and Intermediate distribution frames*

The MDF has all the protective devices like fuse, heat coils and carbon blocks. The fuses protect the wires and the instruments from high currents. The heat coil earths the line if current persists for a long time and carbon blocks protect from a high transient voltage. It is seen that MDF has two terminals A and B which are extended from the MDF to the IDF. However, whereas the MDF has only two wires, the

IDF has three terminals on the left and four to the right for one subscriber. The three terminals go to the three conductors of the multiple jacks and to the subscriber's meter. The four terminals have the circuits of answering jack, calling lamp and cut-off relay of subscriber's line circuit. These terminals of distribution frames are made connected at the time of installation and usually not altered.

1.10 CROSSBAR SWITCH

Since 19th century switching system have used and continue to use a wide variety of electromechanical automatic switch such as stronger switch. At present, most popularly used electromechanical switch, crossbar switch is used. In 1919, two Swedish Engineers, Betulander and Palmgren got patent for crossbar switch. In 1938, AT & T laboratories in US introduced crossbar switching system in the field.

In a crossbar switch, an array of horizontal and vertical wires (solid lines) are connected to initially separated contact points of switches, Horizontal and vertical bars (dashed lines) are mechanically connected to these contact points. The bars (horizontal and vertical) are connected to electromagnets when the bars are activated, contact points connected to the bars move closer to its facing contact points. The motion is not enough to cause contact so that electromagnets connected to both the bars are activated and contact is made. There is mechanical latching incorporated into crossbar switches so that when a horizontal and vertical bars are activated, a contact is made and continues to hold even if the horizontal bar activation is released. The contact will not open until the vertical activation is also released. A basic crossbar switch shown in Fig. 1.7.

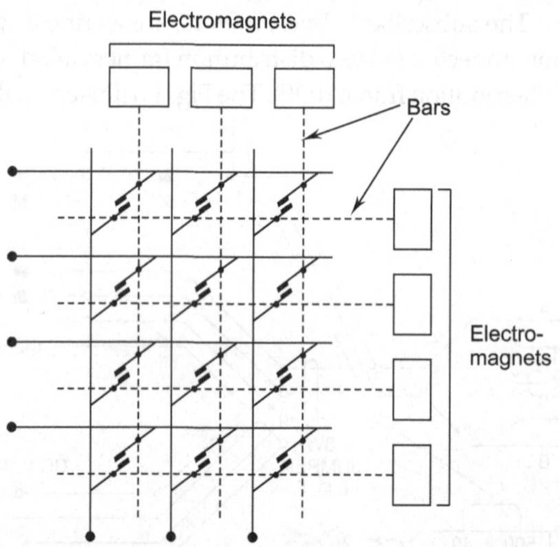

Fig. 1.7 *Illustrating the principle of basic crossbar switch*

Features of Crossbar Switch :

- Flexible concentration ratios allow the system designer to select the appropriate ratio for a specific switch in a specific location.
- Crossbar switches are easier to maintain as there is a few moving parts.
- Crossbar switch is used with common control switching system which allows the switch to share the common equipments used to process the call.
- Wire logic computer allows some particular routine functions of call processing to be handwired into the switch.

Note. Common Control Switching with the help of computers or microprocessors perform event monitoring, call processing charging, operation and maintenance. This is the fundamental concept of crossbar switching.

1.11 A GENERAL TRUNKING

A circuit or channel between two switching system is called 'trunk'. The subscriber all over the world connect each other through switching system. A single switching system can connect subscribers within a few kilometers from the system. So many switching systems of different locations or cities connect each other for global connectivity. A Fig. 1.8 shows the subscriber lines and trunking lines or trunks. A very known example of trunking circuit is Subscriber Trunk Dialling (STD) or Direct Distance Dialling (DDD).

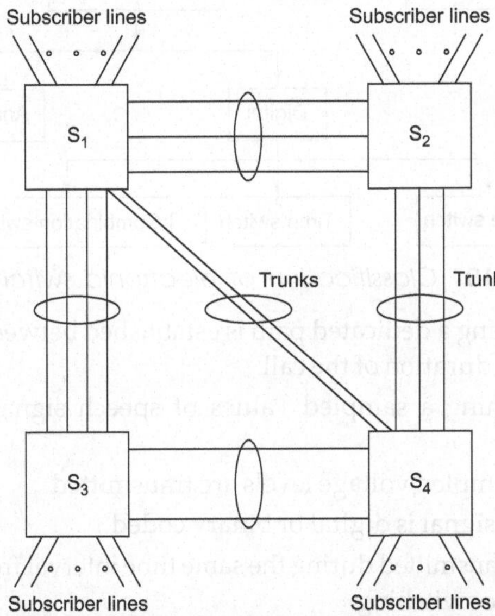

Fig. 1.8 *Trunk lines in the telecommunication network (S_1, S_2, S_3, S_4 are the switching systems)*

Here the number of trunks may vary between pairs of switching system because it is determined on the basis of traffic between them. A universal trunk circuit is shown in Fig. (1.9).

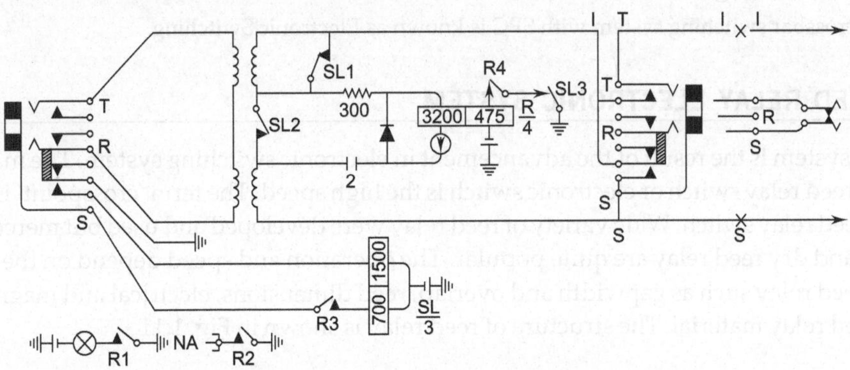

Fig. 1.9 *Universal Trunk Circuit*

1.12 ELECTRONIC SWITCHING

In electronic switching system, the control functions are performed by a computer or a microprocessor. Thus electronic switching also known as stored program control (SPC) system. New advanced technologies or facilities can be added by changing the control program stored as a software. The classification of electronic switching system is shown in Fig. 1.10.

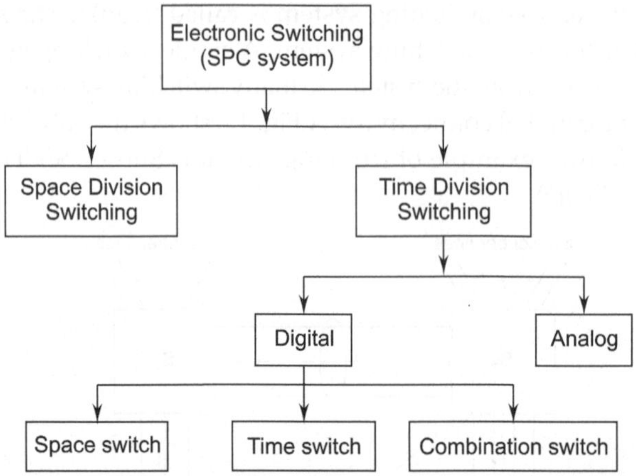

Fig. 1.10 *Classification of electronic switching*

- In space division switching a dedicated path is established between the calling and the called subscriber for the entire duration of the call.
- In time division switching a sampled values of speech signals are transferred at fixed intervals.
- In analog switching a sampled voltage levels are transmitted.
- In digital switching the signal is digital or binary coded.
- If the coded values are transmitted during the same time interval from input to output, then the technique is called space switching.
- If the coded values are stored and transferred to the output at a later time interval, then the technique is known as time switching.
- When the combination of space switching and time switching are designed to use in case of time division digital switch then it is known as combination switch.

Note. A crossbar switching system with SPC is known as Electronic Switching.

1.13 REED RELAY ELECTRONIC SYSTEM

Reed relay system is the result of the advancement in electronic switching system. The main advantage of the reed relay switch or electronic switch is the high speed. The term 'crosspoint' is also used to denote reed relay switch. Wide variety of reed relay were developed and used but mercury welted reed relay and dry reed relay are quite popular. The operation and speed depend on the structural details of reed relay such as gapwidth and overlap, reed dimensions, electrical and magnetic properties of reed relay material. The structure of reed relay is shown in Fig. 1.11.

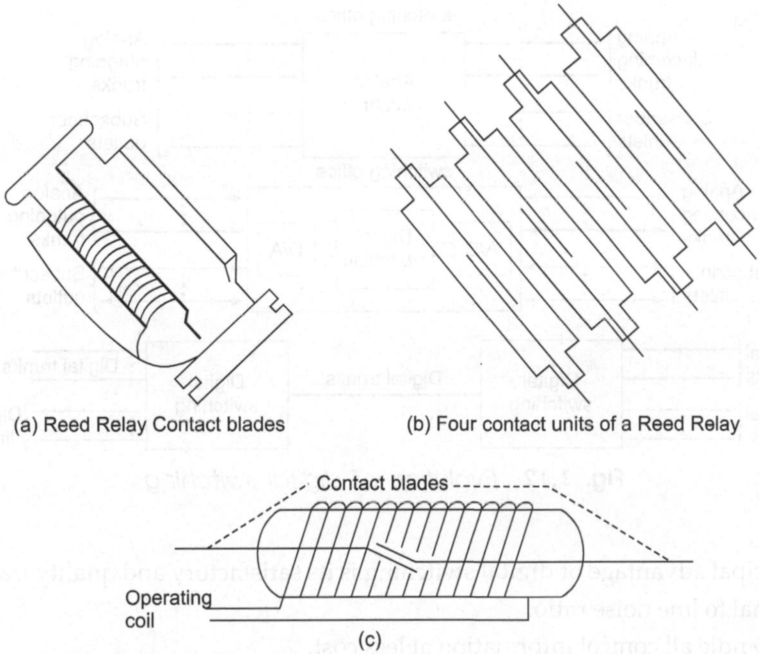

Fig. 1.11 *Detail of a reed relay coil and contact units*

Principle of Working : When an electric current is passed through a coil, surrounded the relay contact unit, generates a magnetic flux through the reed contact. When the flux passes through the contact gap, the two reed blades gets an attractive force and make action results. The contact blades are long and thin and are made of special alloy material. The contact areas of the blades are covered with special contact material and the glass tube surrounded the contact unit is sealed and filled with special gases. The contact blades are so designed and positioned that their diplection produces the required release force.

Applications : The reed relay switches are generally used in analog electronic exchange employing space division switching.

- Reed relay system is used with coaxial shielding for high frequency applications and process control equipments.
- Due to low cost and versatility it is suitable for many security and general purpose applications.
- Some advantages like long life, small size, low capacitance, high insulation resistance and low and stable contact resistance make reed relay to serve in many different applications.

Note. For reed relay switch 2.86 mA current is required and it operates in about 2 msec.

1.14 DIGITAL SWITCHING SYSTEM

When the input to and output from the switching system support digital signal then the switching system is known as digital switching system. In the present day, most of the switching system is digital switching system. Time switching and space switching are used in digital switching system. A simplified diagram of digital switching system is shown in Fig. 1.12.

In the next chapter (*i.e.*, in chapter 2), we will discuss in detail the Digital switching system.

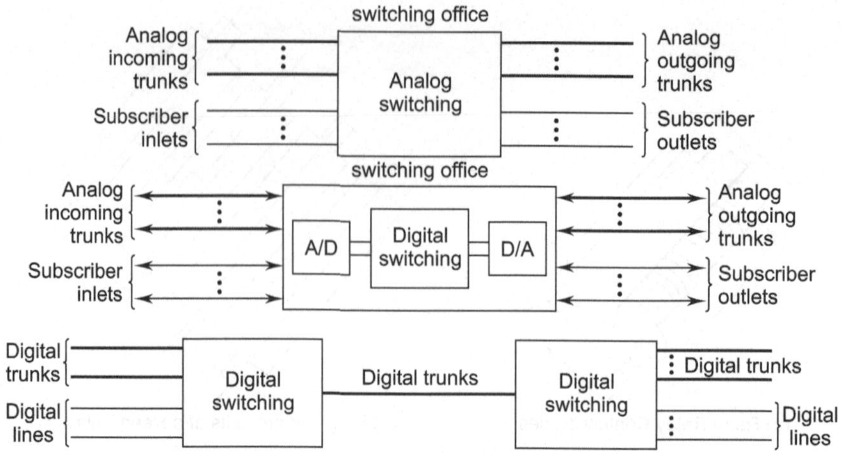

Fig. 1.12 *Evolution of digital switching*

Advantages

1. The principal advantage of digital switching is its satisfactory and quality transmission.
2. Lone signal to line noise ratio.
3. Easy to handle all control information at less cost.
4. Time required for communication is less.
5. Easy to implement new facilities.

Note. The digital switching system also known as Electronic switching system - III generation. Switching process is by time division/digital transmission and pulse code modulation is used in this switching system.

SUMMARY

- The process of transferring information from one place to another by means of cable or wireless media including necessary hardware and software is known as switching.
- Earlier manual switching system are of two types —
  (*i*) The magneto or local battery system and
  (*ii*) The common battery system.
- Message switching and circuit switching comes under the electronic switching system where computer is used.
- Switching system or the exchange is capable of receiving and storing the dialled numbers, translating the exchange identifier into routing digits, and transmitting the routing and the subscriber line identifier digits to the switching network.
- In common control system the call processing is done independently and this concept is used in crossbar switching.
- To remove the problem of mechanical motion of bars in crossbar system, reed relay is used in advanced switching system.
- When the input to and output from the switching system are digital signals then the switching system is known as Digital switching system.
- In present state, all the switching system is based on digital system.

REVIEW QUESTIONS

1. What do you mean by switching system ? Why it is necessary ?
2. Write briefly one technique of data transfer using switching system.
3. What are the basic functions of switching system ?
4. Compare message switching and circuit switching.
5. Discuss crossbar switch.
6. Write down the definition of following terms
   (*i*) Electronic switching
   (*ii*) Distribution frame
   (*iii*) Trunk
   (*iv*) Called subscriber and calling subscriber.
7. Write short note on Reed relay switch. Give advantage.
8. What is digital switching system ?
9. What do you mean by store and forward network ? Write in short.
10. List some benefits of electronic switching.
11. Classify the switching system.
12. What is meant by register - translator sender ?
13. Discuss some advantages and disadvantages of digital switching over analog switching.
14. Briefly discuss the evolution of digital switching.

MULTIPLE CHOICE QUESTIONS

1. The example of electromechanical switching system ________
   (*a*) Crossbar switching system
   (*b*) Reed relay switching system
   (*c*) Magneto switching system
   (*d*) None of the above.
2. Which type of switching is inefficient of transferring long messages ?
   (*a*) Circuit switching
   (*b*) Message switching
   (*c*) Packet switching
   (*d*) None of the above.
3. The another name of message switching is ________
   (*a*) Line switching
   (*b*) Store and forward switching
   (*c*) None.
4. In case of speed of dialling, which type of switch is advantageous ________
   (*a*) Crossbar switch
   (*b*) Stronger switch
   (*c*) Reed relay switch
   (*d*) None.
5. Trunks in a communication network ________
   (*a*) Links that run to the subscriber premises.
   (*b*) Links that run between the switching systems.
   (*c*) None of these.

ANSWERS

1. (*a*) 2. (*b*) 3. (*b*) 4. (*c*) 5. (*b*)

2

Digital Switching

2.1 INTRODUCTION

In present days, the digital system completely change the whole setup of switching system. The reliability and efficiency of digital switching are the main key to replace the analog switching system. Digital switching matrices were first introduced into the U.S. public network in 1976 with AT and T's No. 4 ESS Toll switch.

To connect pairs of channels is the basic function of digital switching network. To achieve this switching two processes called space switching and time switching are used. The processing function of these two switches are described in this chapter and some another digital systems are also discussed. Digital switching system handles only the digital signal. Assuming that the conversion of Analog signal to digital signal is known, we have just refer the digital signal only.

2.2 SWITCHING FUNCTIONS

The switching functions of a switching office is same as described in chapter 1, irrespective of the system whether it is a manual or electronic (analog or digital) switching system. So, the basic function is to setup and release (make and break) connections between transmission channels according to the requirements. There are some switching categories for voice circuit in order to make the connections.

1. **Local or Line to Line Switching.** In this switching, there is a direct connection from the originating loop to a specific terminating loop and each loop must be accessible to every other loop. Example—a PBX.
2. **Transit or Tandom Switching.** This switching requires a path from an originating line to an outgoing line or trunk group. Generally, more than one outgoing call is acceptable, not necessary that every outgoing line should be accessible from every incoming line. Example—Inter office trunk group.
3. **Call Distribution.** This type of switching is implemented with same basic equipment as PBXs but only difference is in the mode of operation as incoming calls can be available to any

available attendant. To distribute the arriving calls among the attendants, a software named Automatic Call Distribution (ACD) is designed. It also requires non-blocking operation. Fig. (2.1) shows local and transit traffic switching.

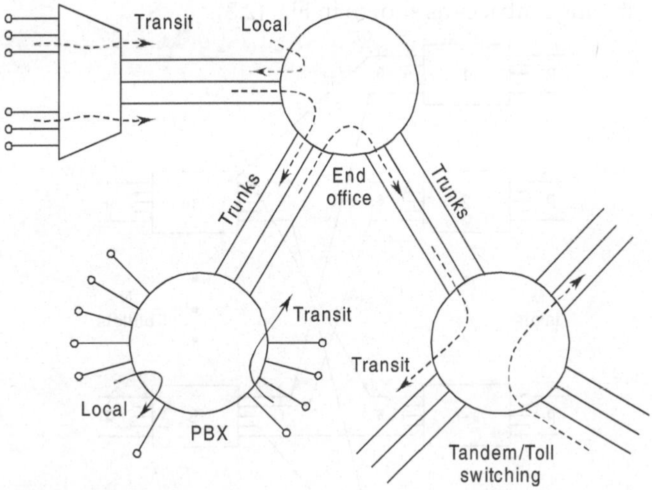

Fig. 2.1 *Illustration of local and transit traffic switching*

2.3 SPACE DIVISION SWITCHING

In space division switching, the paths in the circuit are reparated from each other. Originally, space division switching was designed for analog network but it is used in digital switching as well as analog switching. A cross point switch is known as space division switch as it moves a bit stream from one circuit/bus to another. The switches are hardware and software devices capable of creating temporary connections between two or more subscribers. The number of cross points required are M×N where M is the number of inlets and N is the number of outlets. If available outlets are overlapped for various inlets then this technique is called "grading". The example of grading is rectangular cross point.

2.4 MULTIPLE STAGE SWITCHING

Single stage switching is impractical in now-a-days as number of subscriber is increasing day by day. Therefore multistage network or switching is done. All the disadvantages of single stage switching are overcome by adopting multistage switching.

2.4.1 Two-Stage Network

A block diagram of two stage network is shown in Fig. (2.2). Here any of the N inlets can be connected to any of the K outputs of the first stage. Similarly, any of the K inputs of the second stage can be connected to any of the N outlets. So, there are K alternative paths for any inlet/outlet pair connection.

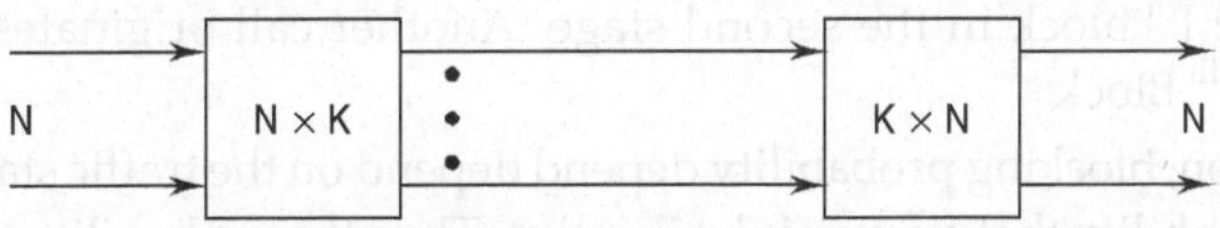

Fig. 2.2 *A two stage n × n network*

This network is said to provide full connectivity or full availability, depending on the concept that any of the N inlets can be connected to any of the N outlets in the network. Now, for large N, it is difficult to realise the switching matrix (N × K). Therefore, we consider an M × N switch using a number of smaller switching matrices as shown in Fig. (2.3).

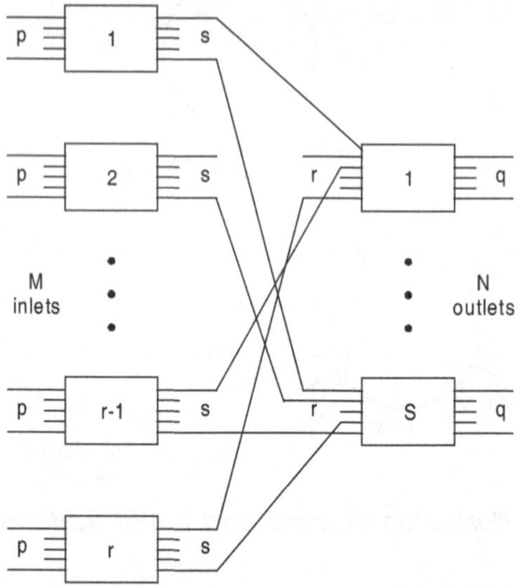

Fig. 2.3 *Two-stage network with multiple switching matrices in each stage*

Here, M inlets are divided into *r* blocks of P inlets *i.e.*, M = *pr*. Likewise N outlets are divided into *s* blocks of *q* outlets *i.e.*, N = *qs*. Under, full availability; there must be one outlet from each block in the first stage terminating as inlet on every block of the second stage. Therefore, the block sizes of first and second stage will be ps and qr respectively. The total number of switching elements S is given by,

$$\boxed{S = psr + qrs} \quad ..(2.1)$$

or we can write,

$$\boxed{S = Ms + Nr} \quad ...(2.2)$$

The switching capacity SC will be given as,

$$\boxed{SC = rs} \quad ...(2.3)$$

i.e., equal to the number of links between first and second stage.

Blocking Probability

The two conditions on which the blocking probability occur are as follows :

(*i*) In the case of uniformly distributed calls, there are $r \times s$ calls in progress and the $(rs + 1)^{th}$ call arrives.

(*ii*) There is no uniformly distribution of calls and there is a call in progress from I^{th} block in the frst stage to the Jth block in the second stage. Another call originates in the Ith block and destined to the Jth Block.

For the first condition, blocking probability depend depend on the traffic statistics. For the second stage, let α be the probability that a given inlet is active. Then the probability that an outlet at the Ith block is active is,

$$\boxed{\beta = \frac{(P\alpha)}{S}} \qquad \ldots(2.4)$$

The probability that another inlet becomes active and search an outlet other than the already active outlet is given by,

$$= \frac{(P-1)\alpha}{(S-1)}$$

Therefore,

$$\boxed{P_B = \frac{P\alpha}{S}\left[1 - \frac{(P-1)\alpha}{S-1}\right]} \qquad \ldots(2.5)$$

Substituting $P = \frac{M}{r}$, we have

$$\boxed{P_B = \frac{M\alpha(S-1) - \left\{\left(\frac{M}{r}\right) - 1\right\}\alpha}{rs(S-1)}} \qquad \ldots(2.6)$$

From the equation (2.6), it is seen that if switching elements reduces then PB goes high. It may be noted that,

if $N > M$, the network is expanding the traffic.
if $N < M$, the network is concentrating the traffic.
if $N = M$, the network need attention.

To improve the blocking performance, the number of links between the two stages should be increased. Let us assume that there are K links between two stages then,

for, $M = N$, the design parameters are

$$P = q = \sqrt{N} \quad \text{and} \quad S = r = K\sqrt{N}$$

Total number of cross points will be $S = 2NK\sqrt{N}$ and switching capacity will be equal to N. To make the network non-blocking, we must have, $K = \sqrt{N}$. Thus, the number of cross points will be $S = 2N^2$ and switching capacity will be equal to N (*i.e.*, SC = N).

Thus, a two stage non-blocking network requires twice than number of switching elements as the single stage non blocking network.

Note. The total number of required cross point switches of a large single stage network is $\frac{N(N-1)}{2}$.

2.4.2 Three-Stage Network

Three stage network is more usefull than two stage network as blocking probability and number of switching elements are less. The figure (2.4) shows the N × N three stage blocking network. Here, the N inlets and N outlets are divided into r blocks of P inlets and P outlets respectively. The sizes of the switching matrices are $P \times s$ in stage 1, $r \times r$ in stage 2 and $s \times P$ in stage 3. In this case, any arbitrary inlet of the first stage has s alternative paths to reach any arbitrary outlet in the third stage.

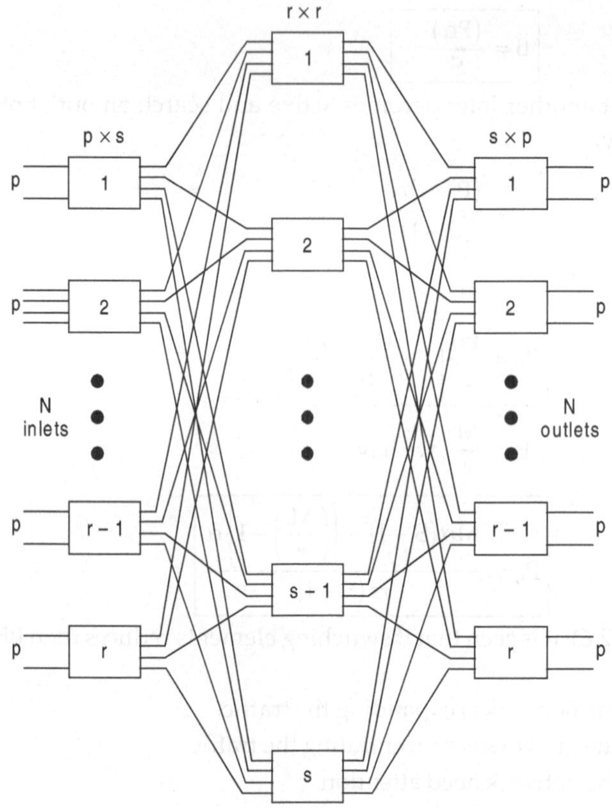

Fig. 2.4 *N × N three-stage switching network*

The total number of switching elements is given by,

$$S = rps + sr^2 + spr = 2Ns + sr^2 = s(2N + r^2) \qquad \ldots(2.7)$$

For square matrices in first and third stages *i.e.,*

$$P = s = \left(\frac{N}{r}\right)$$, the no. of switching elements will be,

$$\boxed{S = \frac{2N^2}{r} + Nr} \qquad \ldots(2.8)$$

By differentiating equation (2.8) with respect to r and equating to zero we have,

$$\frac{dS}{dr} = \frac{-2N^2}{r^2} + N = 0$$

$$\boxed{r = \sqrt{2N}} \qquad \ldots(2.9)$$

and as the 2nd derivative is positive at this value of r, it indicates that S is minimum *i.e.,*

$$\boxed{S_{min} = 2N\sqrt{2N}} \qquad \ldots(2.10)$$

Since, $P = \dfrac{N}{r} = \sqrt{\dfrac{N}{2}}$

The optimum ratio of the number of blocks is to the number of inputs per block is given by,

$$\boxed{\frac{r}{P} = \frac{\sqrt{2N}}{\sqrt{N/2}} = 2} \quad ...(2.11)$$

Blocking Probability

It can be calculated by considering the serial and parallel paths or links.

Let, β = probability that a link is busy.

β^1 = probability that a link is free.

Then $\beta = 1 - \beta^1$

If there are S parallel links the blocking probability is the probability that all the links are busy,

$$P_B = \beta^S, \quad Q_B = 1 - P_B = 1 - \beta^S$$

When a series of *s* links are connected to complete the connection then,

$$P_B = 1 - (\beta^1)^s = 1 - (1 - \beta)S$$

In three stage network, there are two links in series for every path and there are three parallel paths.

Therefore,

$$\boxed{P_B = [1 - (\beta^1)^2]^S = [1 - (1 - \beta)^2]^S} \quad ...(2.12)$$

If α is the probability that an inlet at the first stage is busy then,

$$\boxed{\beta = \frac{P\alpha}{s} = \frac{\alpha}{K}} \quad ...(2.13)$$

Therefore, $$\boxed{P_B = [1 - (1 - \alpha/K)^2]^S} \quad ...(2.14)$$

From equation (2.14), it is seen that the factor K represents either expansion or concentration.

If $s > P$ then first stage provides expansion otherwise it is concentration.

For non-blocking operation, the worst situation occurs when the conditions are as follows :

1. (P – 1) inlets in a block I in the first stage are busy.
2. (P – 1) outlets in a block O in the third stage are busy.
3. The (P – 1) second stage blocks, on which the (P – 1) outlets from block I are terminated, are different from the (P – 1) second stage blocks from which the links are established to the block O.
4. The free inlet of block I needs to be terminated on the free outlet of block O.

Fig. (2.5) illustrated the conditions. An additional block in the second stage needed. Thus the total number of bocks required in the second stage is,

$$\boxed{S = 2(P - 1) + 1 = 2P - 1} \quad ...(2.15)$$

and the no. of swiching elements in the non-blocking system is given by,

$$S = P(2P - 1)r + (2P - 1)r^2 + P(2P - 1)r$$

$$\Rightarrow \quad S = 2N\left(\frac{2N}{r} - 1\right) + r^2\left(\frac{2N}{r} - 1\right) \quad [\because \; P = N/r]$$

$$\boxed{S = \frac{4N^2 - 2Nr}{r} + 2Nr - r^2} \quad ...(2.16)$$

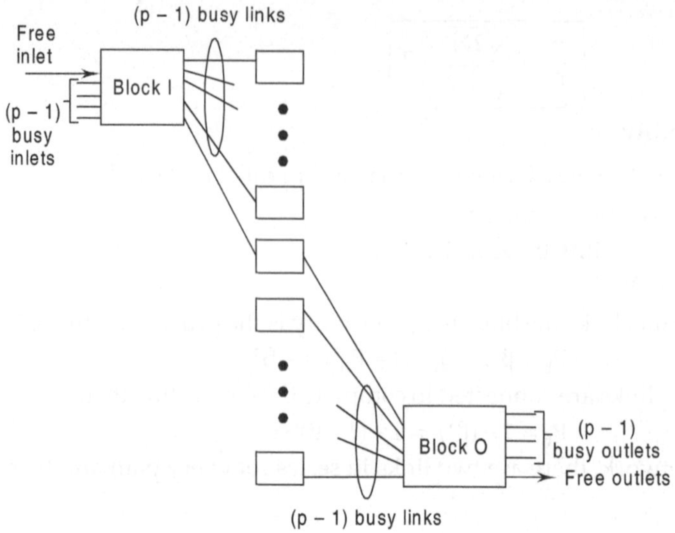

Fig. 2.5 *Three-stage non-blocking configuration*

The size of the switching matrices are P (2P – 1), $r \times r$ and $(2P-1) \times P$ in 1st stage, second stage and third stage respectively. The optimum value of r for minimising the number of switching elements may be calculated as,

$$\frac{dS}{dr} = -\frac{4N^2}{r^2} + 2N - 2r = 0$$

i.e., $\quad -2N^2 + Nr^2 - r^3 = 0$

or, $\quad \boxed{r^2 (N - r) = 2N^2} \qquad \ldots(2.17)$

For large value of N, $\quad N - r = N$ and thus

$$\boxed{r = \sqrt{2N}\,,\ P = \sqrt{\frac{N}{2}}} \qquad \ldots(2.18)$$

Substituting the value of r in equation (2.16) we have,

$$S_{min} = \frac{4N^2}{\sqrt{2N}} - 2N + 2N\sqrt{2N} - 2N$$

$$= 4N\sqrt{2N} - 4N = 4N\left(\sqrt{2N} - 1\right)$$

$$= 4N\sqrt{2N} \qquad \ldots(2.19)$$

The switching elements advantage ratio λ as,

$$\lambda = \frac{\text{no. of switching elements is a non-blocking single stage network}}{\text{no. of switching elements in a non-blocking three stage network}}$$

2.5 NON-BLOCKING SWITCHES

As the single stage switch is non-blocking in nature, same-wise three stage switch may be non-blocking if each individual array is non-blocking and if the no. of center stages (let K) is equal to $2n - 1$. Fig. (2.5) shows the three stage switching matrix which is non-blocking in nature. We have seen that the number of cross-points of a non-blocking three stage switch is

$$\boxed{S_{min} = 4N\left(\sqrt{2N} - 1\right)} \quad \text{...(2.20)}$$

2.6 BLOCKING PROBABILITIES

Every switching system are designed to handle the traffic in the busiest hour of the day. Therefore, it is necessary to study the maximum blocking probability of the system.

It is one of the aspects of GOS of the telephone industries. Depending on accuracy, required availability, complexity, priority, geographical area and applicability of the different network structure, there are various techniques. Among all the techniques two most important techniques are discussed here. The two techniques are :

1. Lee graphs
2. Jacobaeus method.

Lee graph method is most versatile and straight forward approach but Jacobaeus method is more accurate than the Lee graph. We will discuss the two techniques one by one.

2.6.1 Lee Graphs

This is the straight forward approach of calculating probabilities with the use of probability graphs. C.Y. Lee first proposed this methods and that's why named Lee graph.

Let P be the probability that a link is busy and the probability that a link is idle is represented as $q = 1 - P$. The blocking probability B is the probability that all the links are busy if there is 'n' parallel links and any one link is used to complete a connection, then it is given by,

$$\boxed{B = P^n = 1 - q^n} \quad \text{...(2.21)}$$

Now, for a three stage network, the probability of blocking (graph is shown in Fig. (2.6), is given by,

$$\boxed{B = (1 - q'^2)^K} \quad \text{...(2.22)}$$

where, q' = probability that an inter stage link is idle = $1 - P'$

P' = probability that any particular inter stage link is busy.

K = No. of centre stage arrays.

If P is known, the probability that an inter stage link is busy is given by,

$$\boxed{P' = \frac{P}{\beta}} \quad \text{...(2.23)}$$

where $\beta = \dfrac{K}{n}$

= factor by which the percentage of inter stage links that are busy is reduce

Therefore, $$q' = 1 - \frac{P}{\beta}$$ ...(2.24)

and $$B = \left[1 - \left(1 - \frac{P}{\beta}\right)^2\right]^k$$ ...(2.25)

The equation (2.25) gives the complete expression for the blocking probability of a three stage switch in terms of P.

The benefits of this method are :

(*i*) All the formulas are related to network structure.

(*ii*) It provide accurate results

(*iii*) This method provide ideas to change the structure for high performance.

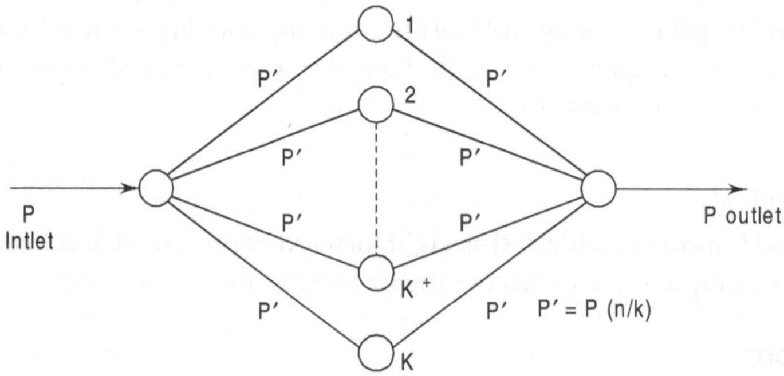

Fig. 2.6 *Lee's graph for a three stage network*

2.6.2 Jacobaeus Method

When space expansion takes place, the assumption of individual independent probability leads to an erroneously high value of blocking. The inaccuracy results because more and more paths in a switch are found to be busy and remaining paths are less to be in use. According to C. Jacobaeus, the blocking probability of a three stage switch is given as,

$$B = \frac{(n!)^2}{K!(2n-k)!} P^K (2-P)^{2n-K}$$ ...(2.26)

where, n = number of inlets/outlets per first/third stage array.

K = number of second sage array.

P = inlet utilization.

Though it is more accurate, but high blocking probabilities have not much practical value. Therefore use of this technique is not considered here.

2.7 COMPARISON OF SINGLE STAGE AND MULTISTAGE NETWORKS

Single stage	*Multistage*
1. Inlet to outlet connection is through a single cross point.	1. Inlet to outlet connection is through multiple cross points.
2. A particular cross point is required for each particular connection.	2. A particular connection may be established by using different sets of cross points.
3. Each individual cross point can be used for only one.	3. Same cross point can be used to establish connection between a number of inlet/outlet pair.
4. Use of single cross point per connection results is better quality link.	4. Use of multiple cross points may degrade the quality of a connection.
5. If a cross point fails, no associated connection will be established as there is no redundancy.	5. Alternative cross points and paths are available.
6. Cross points are inefficiently used. Only one cross point in each row/column of a square/triangular switch matrix is ever in use, even if all the lines are active.	6. Cross points are used efficiently.
7. Number of cross points is prohibitive.	7. Number of cross points is reduced significantly.
8. A large number of cross points in each inlet/outlet leads to capacitive loading.	8. There is no capacitive loading problem.
9. The network is non-blocking in nature.	9. The network is blocking in nature.
10. Time for establishing a call is less.	10. Time for establishing a call is more.

2.8 FOLDED FOUR-WIRE SWITCHES

Four-wire switching operation is used by multiple stage switches where two paths must be established to complete connection. The Fig. (2.7) shows a four-wire connection through a four-stage switch as there is mirror image effects.

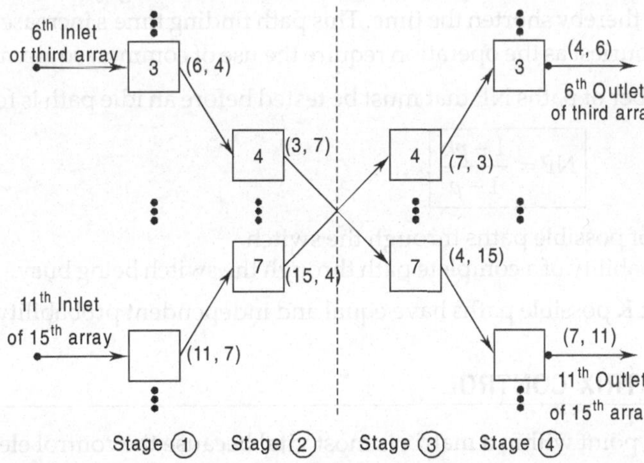

Fig. 2.7 *Four wire connection through Four-stage switch*

If the diagram of the switch is folded about the vertical center line, the paths coincide. This method of setting up connection is known as folded operation with four wire switches. There are several advantages of this folded operation.

Advantages

1. Only one path finding operation is necessary as the reverse path is a mirror image of the forward path. It means that every cross point of an array can make pair with another cross point of corresponding array of opposite side *i.e.*, cross-point (i, j) in one array make pair with cross point (j, i) in the corresponding array of opposite side. Once the forwared path or one cross point in a pair is available then automatically reverse path or another cross point of the pair is specified and available.
2. The total amount of status information of a switch becomes half.
3. The design of foulded four wire structure is obtained by blocking probability which is one half of the probability of finding two paths independently. The pairing makes the paths available for a particular connection in a strict manner.

As we have discussed the even stages of folded four wire switch, the odd stages are also used. In that case, there is a centre stage that contain even no. of arrays and is folded about a horizontal line at the centre of he stage.

2.9 PATH FINDING

Selection of path in single stage switch is virtually automatic because the required cross point is uniquely specified by inlet-outlet pair to be connected. But in case of multiple stage switch it is complicated to avail more than one path. So, that call processor always keep eye that which paths for a particular connection are available in a state store and for this a path finding routing helps to process the state store information and select an available path. The path finding routing also update the state store information whenever a new connection is made and old connection is released.

Path Finding Times: Path finding time *i.e.*, to find an available path, the required time is directly dependent on the number of potential paths checked before any idle path is found. In other way, utilization of link gives the number of potential paths. Some switching system use parallel process to test a no. of paths and thereby shorten the time. This path finding time's increase when the common control equipment is buriest as the operation require the use of common equipment.

The expected number of paths NP that must be tested before an idle path is found is given as,

$$\boxed{\text{NP} = \frac{1 - pk}{1 - p}} \qquad \text{...(2.27)}$$

where, K = no. of possible paths through the switch.
P = probability of a complete path through the switch being busy.

In this case, each of K possible paths have equal and independent probability of being busy.

2.10 SWITCH MATRIX CONTROL

The selection of cross point within a matrix is most vital because the control element of the switch transfer all the information to the network by connecting the proper cross points. There are two methods to select the cross points within a matrix. They are :

(*i*) Output associate control

(*ii*) Input associated control.

In case of output associated control, all the control information may be associated with output lines and specify the inputs to be connected to the corresponding outputs.

In the second method, the control information may be associated with each input and specify the outputs to the corresponding inputs are to be connected. The illustration of these two methods are shown in Fig. (2.8).

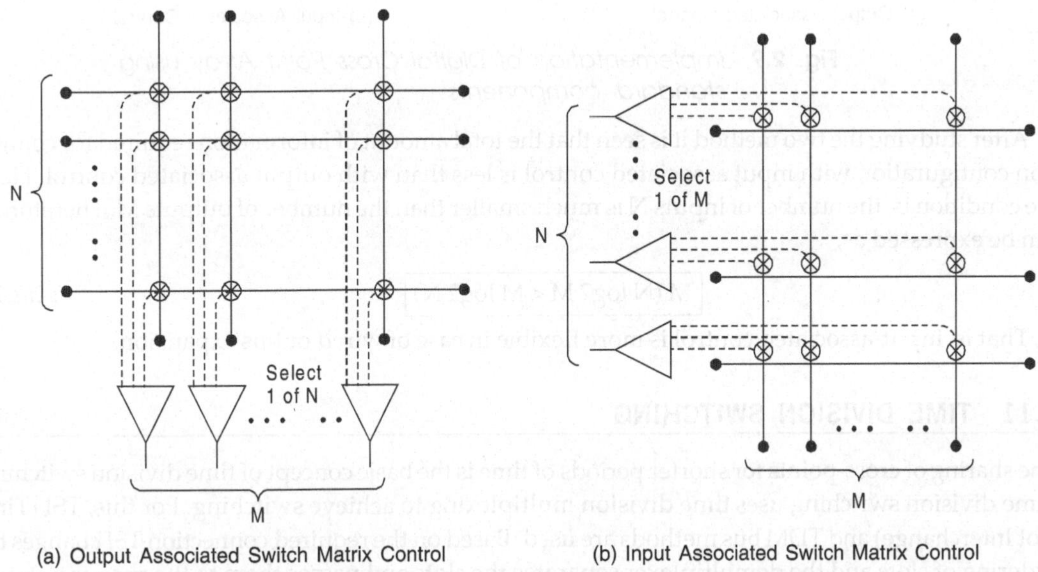

Fig. 2.8 *Two methods of switch matrix control*

Step by step switches are used for input associated control where all the information arrive on the input link and select the O/P link of each stage. In common control system, the source and destination line address is available. Thus, while selecting input to each stage, the connection can be established by beginning at the desired outlet and proceeding backward through the switch. It can be implemented by using conventional live decoders/demultiplexers. The output uses 'wired' or logic function so that output gates of each decoder circuit must be open collector or tri-state devices if TTL (Transistor-transistor logic) is used. In this care the required number of bits to specify a connection configuration is N log2 (M), where N is the number of inlets. One disadvantage of this method is that unused inputs may cross connects when another input alredy select the same output.

The output-associated control uses a conventional data selector/multiplexer for each matrix output. Here, unused outputs can remain connected to an input without allowing that input from being selected by another output. The total number of bits required to make the complete connection is M log N. In other words we can say that in digital switching network, the output associated control obtain the greater speeds of operation. The implementation of digital cross point array using standard components are shown in Fig. (2.9).

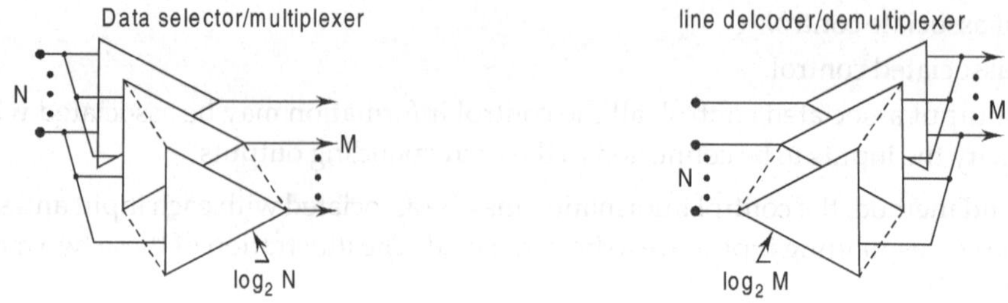

(a) Output Associated Control

(b) Input Associated Control

Fig. 2.9 *Implementation of Digital Cross Point Array using standard components*

After studying the two method it is seen that the total amount of information required for connection configuration with input associated control is less than with output associated control. Here, the condition is, the number of inputs N is much smaller than the number of outputs M. Therefore, it can be expressed as,

$$\boxed{M\,(N \log2 M < M \log2 N)} \qquad \ldots(2.28)$$

That is, input-associated control is more flexible in case of wired or bus expansion.

2.11 TIME DIVISION SWITCHING

The sharing of cross-points for shorter periods of time is the basic concept of time division switching. Time division switching uses time division multiplexing to achieve switching. For this, TSI (Time Slot Interchange) and TDM bus methods are used. Based on the required connection TSI changes the ordering of slots and the demultiplexer separates the slots and passes them to the proper outputs. TDM contains a control unit that opens and closes, the gates according to the switching requirement.

PAM (Pulse Amplitude Modulation) and PCM (Pulse Code Modulation) are used for speech values in digital transmission. With 8 KHz sampling rate, a sample occurs every 125 microseconds. This sample value can be passed from an inlet to an outlet in a few microseconds or less through the cross point. Therefore, during 125 ms sampling interval, a dedicated cross point remains idle for most of the time (say 120 μs). By establishing the dynamic control mechanism, a single cross point can be used to transmit speech samples from a number of inlets to the corresponding outlets. This is the concept of sharing of cross point, the basic principle of time division switching.

Time division switching is equally applied to analog and digital switching system. Due to noise, cross talk and distortion analog switching in used in smaller switching system. Therefore, time division switching is divided into

(*i*) analog time division switching
(*ii*) digital time division switching.

2.12 ANALOG TIME DIVISION SWITCHING

When PAM samples are switched in time division manner, the switching is known as Analog Time Division Switching. Here, a single switching bus supports multiple number of connections *i.e.*, from receive line interface to transmit line interface. There are two cyclic control stores one in the input section and other in the output section, keep the synchronization between the two. The first one controls the receive of input one sample at a time onto the bus and the second one selects the corresponding output line for each input sample. Fig. (2.10) illustrate the working of analog time division switching system structure.

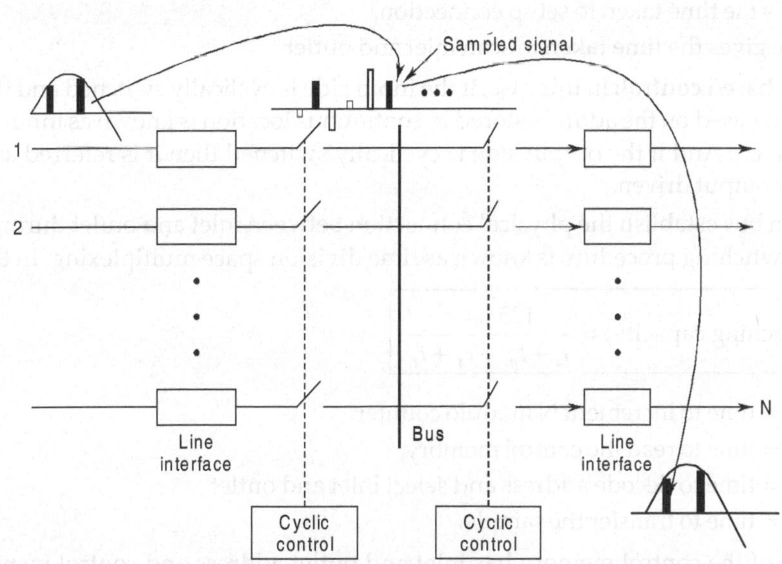

Fig. 2.10 *Structure of Analog Time Division switching*

In analog time division switching, the selection of inlet/outlet is controlled by :

(*a*) Cyclic control (*b*) Memory based control

(*a*) **Cyclic control:** The Fig. (2.11) shows the cyclic control using Modulo-N counter and K to 2^K decoder.

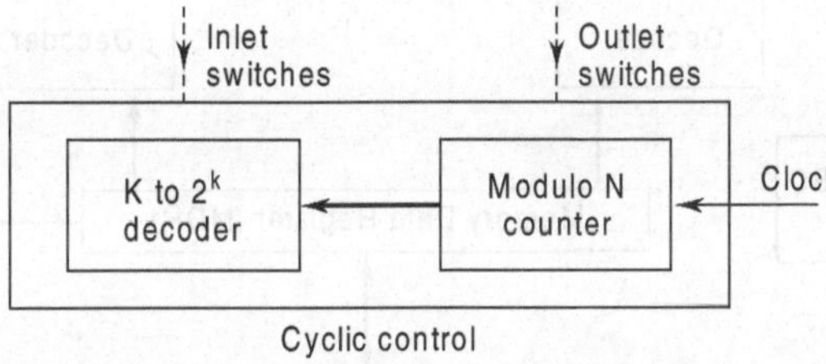

Fig. 2.11 *Cyclic control structure*

Here, K and N are related by,

$$\boxed{\lceil \log_2 N \rceil = k} \qquad \ldots(2.29)$$

where,

N = no. of inlets/outlets

k = decoder size

$\lceil\ \rceil$ = gives the lowest integer.

The switching or channel capacity of cyclic controlled system is given as,

$$C = \frac{125\ \mu\text{sec}}{t_s} \qquad \ldots(2.30)$$

where, it's the time taken to setup connection.

125 m sec gives the time taken to scan inlet and outlet.

(*b*) **Memory based control:** In this case, if the input side is cyclically switched and the outlets are connected based on the address stored in continuous location is known as input controlled or input driven. And if the output side is cyclically switched then it is referred as output controlled or output driven.

The common bus establish the physical connection between inlet and outlet during one sample transfer. This switching procedure is known as time division space multiplexing. In this system,

$$C\ (\text{switching capacity}) = \frac{125\ \mu\text{sec}}{t_i + t_m + t_d + t_t} \qquad \ldots(2.31)$$

where, t_i = time to increment N modulo counter

t_m = time to read the control memory

t_d = time to decode address and select inlet and outlet

t_t = time to transfer the sample.

Every word of the control memory has inlet and outlet address and control memory width is $2\lceil \log_2 N \rfloor$. Reading is done one after one and the modulo counter is updated at the clock rate. For example, between k^{th} inlet and j^{th} outlet path will be established, so that the addresses are entered into the control memory and the location in marked busy as path setup is done. When conversation is over, the addresses are replaced by null values and location is marked free. This is illustrated in Fig. (2.12).

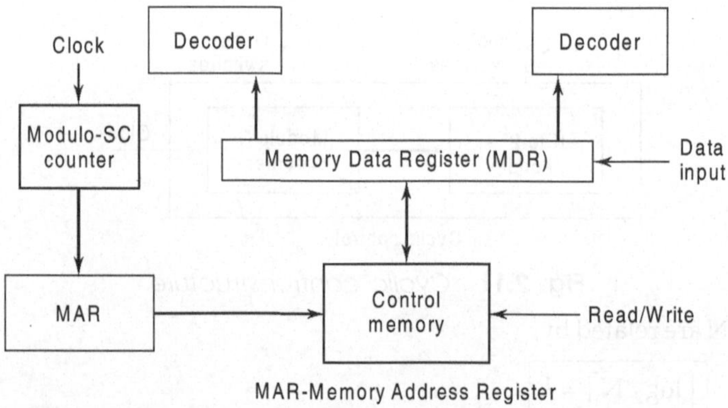

Fig. 2.12 *Memory control for both inlet and outlet*

2.13 DIGITAL TIME DIVISION SWITCHING

If PCM binary samples are switched in a time division manner, then the switching is called Digital Time Division Switching. In this switching, time division switch can be controlled by three ways *i.e.*,

TSI control, sequential write/random read or random write/sequential read control and TDM bus control. The basic functional diagram of Digital time division switching is shown in Fig. (2.13).

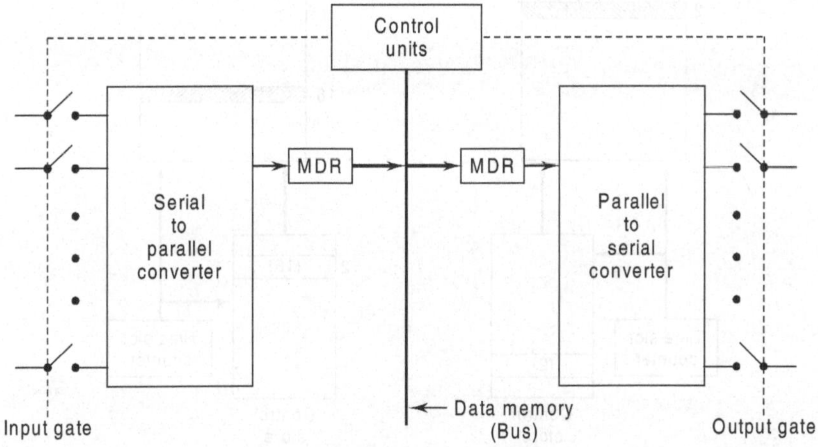

Fig. 2.13 *Basic Functional diagram of Digital time division switching*

A time switch operates by writing data into and reading data out of a single memory. So, in this process, the information in selected time slots is interchanged as shown in Fig. (2.14). The RAM contain several memory locations, each size is same as of single time slot. For data write and read, serial to parallel and parallel to serial converters are used. MDR's are used to connect inlet/outlet. Through input and output gates, the input and output lines are connected to the high speed bus. This gates are used to transfer data from one input line to a specific output line through the bus. The control unit opens and closes the gates according to switching need.

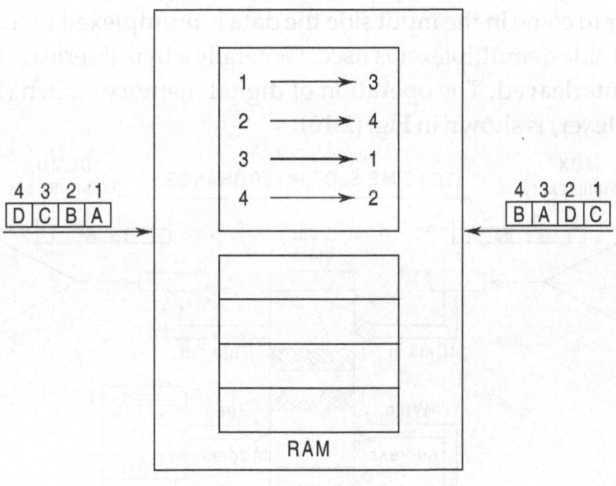

Fig. 2.14 *The operation of TSI*

Another way of controlling *i.e.*, sequential write/random read or random write/sequential read is illustrated in Fig. (2.15).

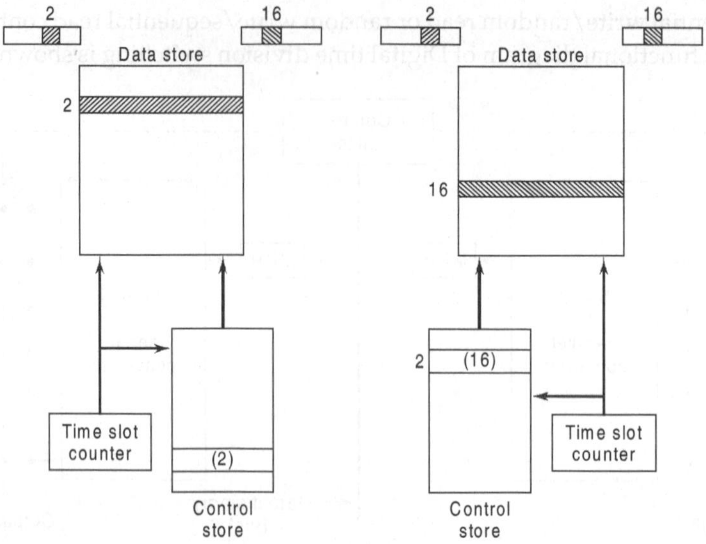

Fig. 2.15 *(a) Sequential write/random read (b) Random write/sequential read*

In case of random write/sequential read, the incoming data are written into memory specified by the control/store and outgoing data are retrieved sequentially under control of an outgoing time slot counter.

The other way of control is the TDM link. The specific memory locations are dedicated to respective channels of the incoming TDM link.

2.14 A DIGITAL MEMORY SWITCH

In digital memory switch data write/read is done sequentially or randomly in RAM using TSI and TDM link control. Prior to come in the input side the data is multiplexed for each direction of travel. Samewise at the output side demultiplexer is used. Generally a byte interleaved form of multiplexing is used instead of bit interleaved. The operation of digital memory switch (Multiplexer-Time slot interchange-Demultiplexer) is shown in Fig. (2.16).

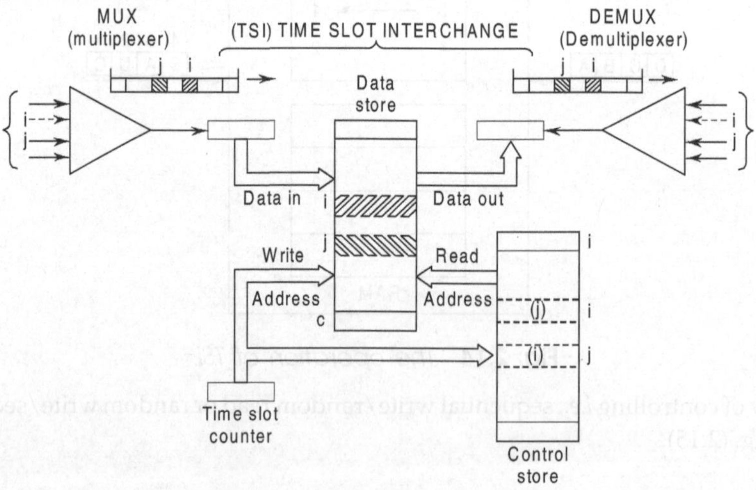

Fig. 2.16 *Digital Memory switch (MUX/TSI/DEMUX) Memory switch*

Here, data store memory is accessed twice during each link time slot. First, some control circuitry selects the time slot no. as a write address and secondly, the content of control store for that particular time slot is selected as a read address. The maximum no. of channels no. that can be supported by the simple memory switch is given as,

$$\boxed{NC = \frac{125}{2t_c}\ \mu\text{sec}} \qquad ...(2.32)$$

where 125 is the frame size for 8 KHz sampled voice signal and t_c is the memory cycle time.

Note. A complete set of pulses, arriving at each active input line is referred as frame.

2.15 TIME STAGES IN GENERAL

Delay is an important factor in case of proper TSI and it can be easily implemented by using RAMs that are written into as data arrive and lead from when data are to be transferred out. The information for each TDM channel can be stored for upto one full frame time without being overwritten. Generally there are two modes of operation that control the time stage memories :

(*i*) Sequential Write/Random Read

(*ii*) Random write/Sequential Read

Both the modes of operation is discussed in section (2.13) with Fig. (2.15). In case of multiple stage design, use of one mode of operation in one time stage and other mode of o-operation in another time stage is possible.

2.16 TWO DIMENSIONAL SWITCHING

Two dimensional (2D) switching is the combination of time and space switch. For a given technology, this switching configuration permit a large number of simultaneous connections. Here both the time slot interchange and sample switching across trunks are achieved. A two stage combination switch *i.e.,* time switch as first stage and space switch as the second stage (TS switch) or vice versa (ST switch). Likewise, three stage combination switches are TST, STS and typical configurations are TSST, TSSSST and TSTSTS TS etc, are discussed in next sections. A general block diagram of combination or two dimensional digital switch is shown in Fig. (2.17).

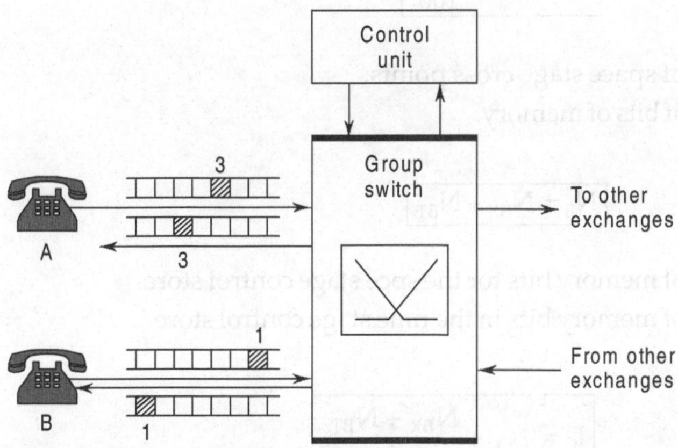

Fig. 2.17 *General block diagram of combined switching*

From the Fig. (2.17) it is seen that the subscriber A makes a call to the subscriber B and for that control unit has assigned time slot 3 to the call on its way into the group switch and slot 1 on its way out of the group switch (to B). During the total conversation this process is maintained. Similarly B to A also carried out. The diagram of TS is shown in Fig. (2.18).

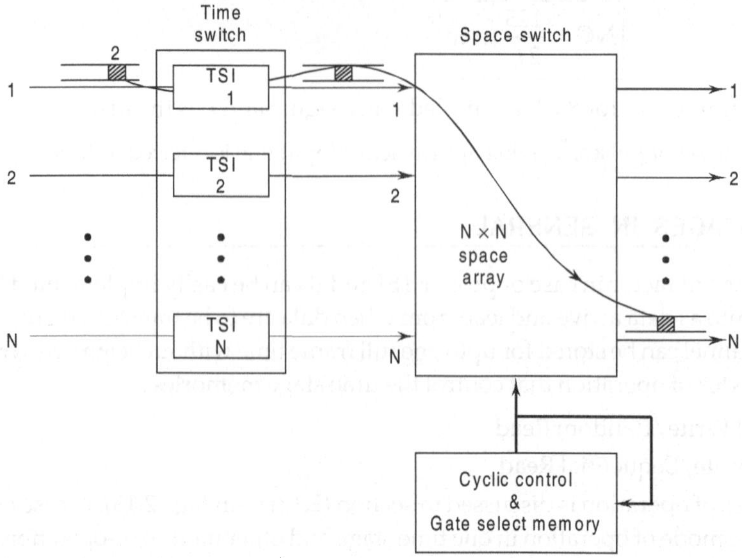

Fig. 2.18 *Time Space (TS) switching*

2.17 IMPLEMENTATION COMPLEXITY OF TD SWITCHES

Basically the complexity of implementation of the switching is represented in terms of no. of cross points and its associated cost. In space stage, the no. of cross points based on the array size and the cost of cross point is relative to the cost of a bit of memory in time stage. To take this into account the cost of memory bit is assumed one hundredth of the cost of cross point. Therefore, the implementation complexity is given as,

$$I_C = N_X + \frac{N_B}{100} \quad \ldots(2.33)$$

where,

N_X = no. of space stage cross points.

N_B = no. of bits of memory.

Now,

$$N_B = N_{BX} + N_{BT}$$

where,

N_{BX} = no. of memory bits for the spce stage control store

N_{BT} = no. of memory bits in the time stage control store.

Therefore,

$$I_C = N_X + \frac{N_{BX} + N_{BT}}{100} \quad \ldots(2.34)$$

2.18 MULTIPLE STAGE TIME AND SPACE SWITCHING

In time division switching, time stage switching is less expansive than space stage switching as digital memory is much cheaper than digital cross points. So in practical, more than one channels are multiplexed together and perform as much switching in time stages to reduces the cost. For reduction of cost multiple stages are used like TST, STS, TSSST etc.

These structures also permits a large number of simultaneous connections to be supported by a given technology. The coming sections describe all the configuration of the switches.

2.18.1 STS Switching

A space-time-space (STS) switching consist of an N × K space matrix at the input, an array of K TSI switches in middle and a K × N space matrix at the output *i.e.*, the time stage is sandwiched between two space arrays. The figure (2.19) shows a simple STS switch where the choice of input and output time slots is fixed for a given connection. But there is a provision of flexibility to use any free TSI switch by space switching on the input and the output side. For this many alternative paths are provided. The nature of this switching is non-blocking.

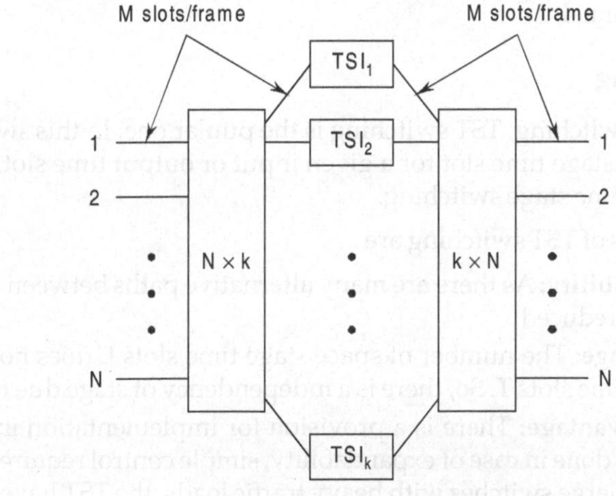

Fig. 2.19 *A STS switch.*

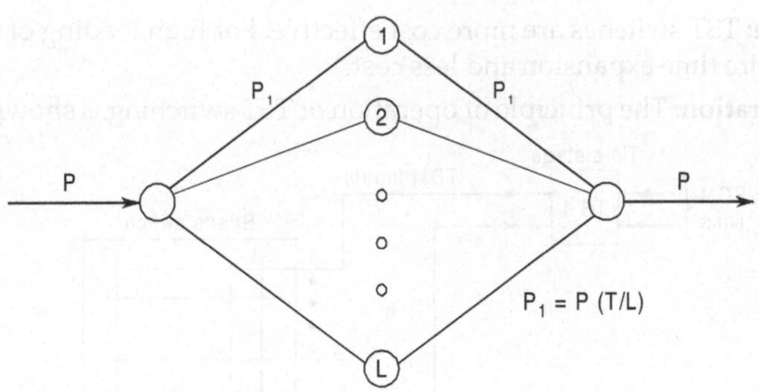

Fig. 2.20 *Probability graph of a 3-stage space switch*

Blocking Probability: The blocking probability of STS switch is given by,

$$B = \left[1 - \left(1 - \frac{p}{\beta}\right)^2\right]^k \qquad \ldots(2.35)$$

where P = Probability that a link is busy

$\beta = \frac{k}{N}$ = a factor by which percentage of probability that a link is busy is reduced.

$(\beta < 1)$

K = no. of centre stage TSI.

The probability graph of STS switch is similar to that of 3-stage space switch (figure 2.20).

Implementation Complexity

$$IC = 2KN + \frac{2KC \log_2 N + KC\,(8) + KC \log_2 C}{100} \qquad \ldots(2.36)$$

where, K = The minimum no. of centre stage TSI to provide desired grade of service

C = No. of channel,

2.18.2 TST Switching

Within all the multistage switching, TST switching is the puular one. In this switching, there is no need to have a fixed space stage time slot for a given input or output time slot. The space stage is sandwiched between two time stage switching.

Some important features of TST switching are :

(*i*) **Low blocking probability:** As there are many alternative paths between two subscribers, so blocking probability reduced.

(*ii*) **Independency of stage:** The number of space stage time slots L does not coincide with the number of external time slots T. So, there is a independency of stage due to TDM links.

(*iii*) **Implementation advantage:** There is a provision for implementation in the TST switches. Implementations are done in case of expandability, simple control requirements, traffic loads etc. It is seen that for large switches with heavy traffic loads, the TST have good implementation advantage.

(*iv*) **Cost effective:** TST switches are more cost effective. For high loading of input channel, TST switches require time expansion and less cost.

Principle of Operation: The principle of operation of TST switching is shown in Fig (2.21).

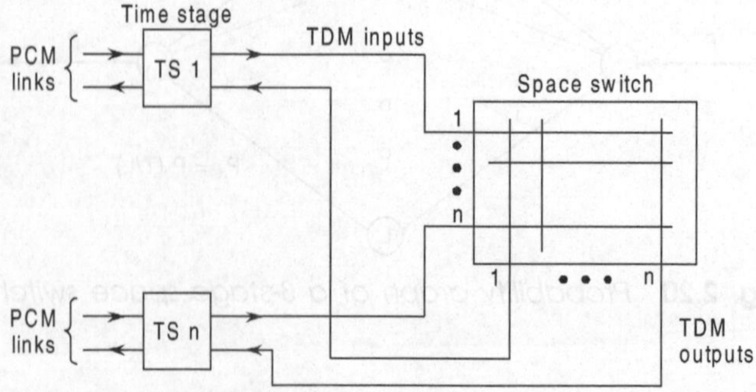

Fig. 2.21 *Principle operation of TST switching*

The two flows of time slots, one for each direction are connected together. Here the information arriving at the incoming link of TDM channel is delayed in the inlet times stage until an appropriate path through the space stage is available. Then the information is transferred to the appropriate outlet time stage through the space stage.

Blocking Probability: Generally, a TST switch is identical to a three-stage space division network. By direct analogy of three-stage space switches, the TST switch is non-blocking if

$$\boxed{L = 2T - 1} \qquad \ldots(2.37)$$

where L = no. of space slot of space switch

T = no. of time slot of time switch

The equivalent lee's probability graph of TST switch is shown in Fig. (2.22) and (2.23)

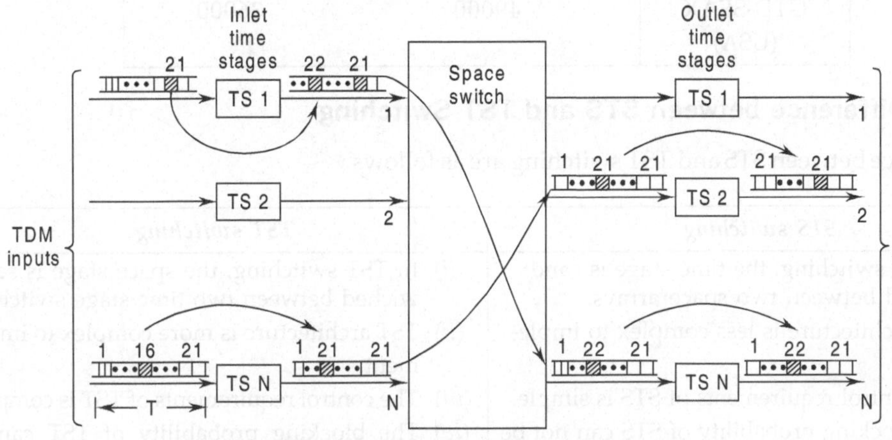

Fig. 2.22 *TST switching structure*

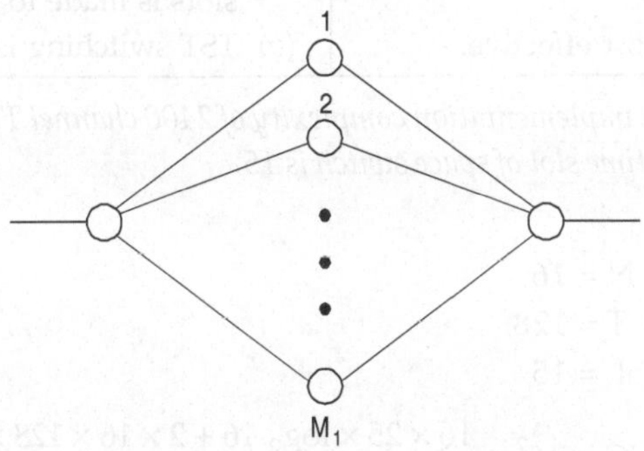

Fig. 2.23 *Probability graph*

The general expression of blocking probability for a TST switch with non-blocking individual stage is,

$$\boxed{B = \left[1 - \left(1 - \frac{PT}{L}\right)^2\right]^L} \qquad \ldots(2.38)$$

Implementation Complexity: The implementation complexity of a TST switch is given by,

$$IC = N^2 + \frac{NL \log_2 N + 2NT(8) + 2NL \log_2 T}{100} \quad ...(2.39)$$

Some popular digital switching systems using TST are tabulated in table below.

Type	Characteristics	
	Maximum no. of trunks	**Traffic (Erlangs)**
E 10 B (France) 1970	3600	1600
C-DOT MAX-XL (India)	40000	47000
GTD-SEAX (USA)	49000	36000

2.18.3 Difference between STS and TST Switching

The difference between STS and TST switching are as follows :

STS switching	*TST switching*
(*i*) In STS switching, the time stage is sandwiched between two space arrays.	(*i*) In TST switching, the space stage is sandwiched between two time stage switches.
(*ii*) STS architecture is less complex to implement.	(*ii*) TST architecture is more complex to implement.
(*iii*) The control requirements of STS is simple.	(*iii*) The control requirements of TST is complex.
(*iv*) The blocking probability of STS can not be minimized.	(*iv*) The blocking probability of TST can be minimized if the no. of space stage time slots is made to be large.
(*v*) STS switching is less cost effective.	(*v*) TST switching is more cost effective.

Example 2.1 *Determine the implementation complexity of 2400 channel TST switch with 16 TDM links and 128 channels. Assume the time slot of space switch is 15.*

Solution.

Given,

$$N = 16$$
$$T = 128$$
$$L = 15$$

Now,

$$IC = 16^2 + \frac{16 \times 25 \times \log_2 16 + 2 \times 16 \times 128 \times 8 + 2 \times 16 \times 15 \times \log_2 28}{100}$$

2.18.4 TSSST Switches

When the space stage of a TST switch is large enough then additional control is required. But it is too complex to handle. So, multiple space stages are used and that makes TSSST switches as it contains three space stages. The main advantage of this TSSS switch is less no. of cross points are used.

If the space stage are increased the TSSSST, TSSSSST switches are made. But generally TSSST switch is used.

2.19 NO. 4 ESS TOLL SWITCH

This switch is known as TSSSST switch as it is Time-Space-Time (TST) switch with four stage of space switch. The basic structure of this switch is shown in Fig. (2.24).

From the figure it is seen that the input is 120 channel TDM links formed by multiplexing five DS1 signals together. All the inputs are passed through a decorrelator (to distribute the incoming TDM links or trunk groups across multiple TDM links into the switch matrix. This TDM trunk groups are not independently busy. If the channels of the trunk group are not decorrelated, there will be blocking probability as they all remain in the same path. So, the decorrelator provide alternate paths by space expansion (7 to 8) and time expansion 120 to 128).

The maximum size of the No. 4 ESS toll switch uses 128 decorrelators with seven 120 channel inputs each.

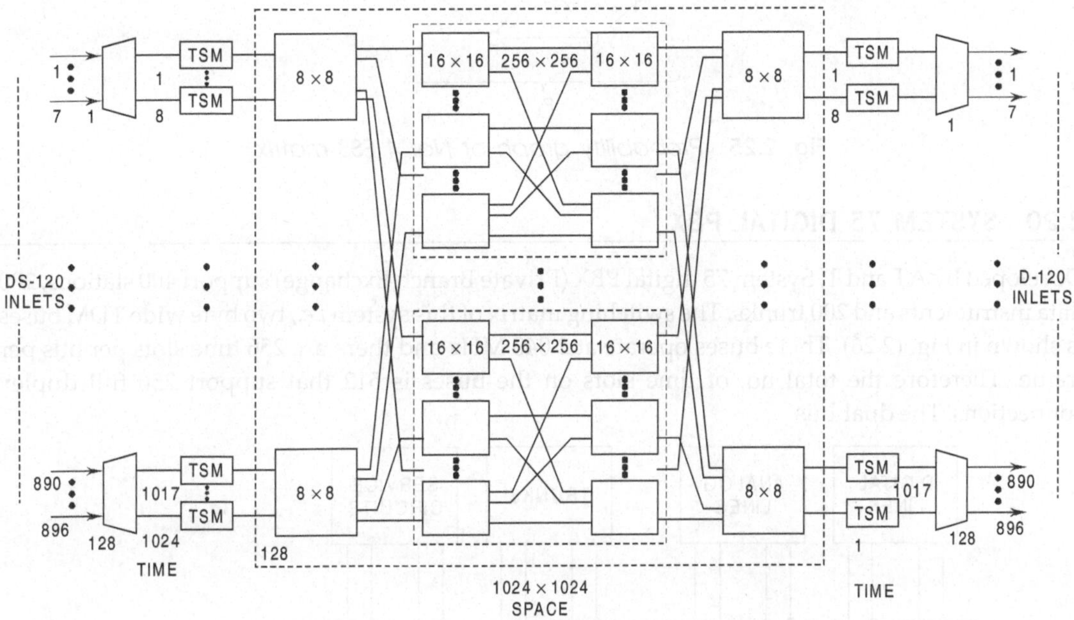

Fig. 2.24 *No. 4 ESS matrix architecture*

Therefore the channel capacity = No. of correlators × 7 × No. of input channels

$$= 128 \times 7 \times 120$$

$$= 107520 \text{ channels.}$$

The blocking probability graph of No. 4 ESS switch is shown in Fig. (2.25).

$$B = \left[1-(1-P_1)(1-P_2)(1-P_1)\right]^{128} \quad \text{...(2.40)}$$

where, P_1 and P_2 is the blocking probability of the four stage space switch.

P is the external channel utilization.

$$P_1 = \left(\frac{7}{8}\right)\left(\frac{120}{128}\right)P \quad \text{...(2.41)}$$

and
$$P_2 = 2p_1^2 q_1^6 + 16p_1^3 q_1^5 + 50p_1^4 q_1^4 + 52p_1^5 q_1^3 + 28p_1^6 q_1^2 + 8p_1^7 q_1 + p_1^8 \quad ...(2.42)$$

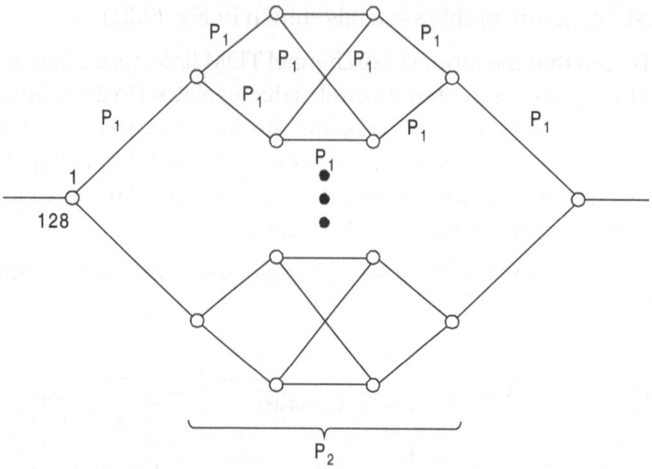

Fig. 2.25 *Probability graph of No. 4 ESS matrix*

2.20 SYSTEM 75 DIGITAL PBX

Developed by AT and T, System 75 digital PBX (Private Branch Exchange) support 400 stations, 200 data instruments and 200 trunks. The switching matrix of this system *i.e.*, two byte wide TDM buses is shown in Fig. (2.26). There buses operate at 2.048 MHz and there are 256 time slots per bus per frame. Therefore the total no. of time slots on the buses is 512 that support 256 full duplex connections. The dual-bus

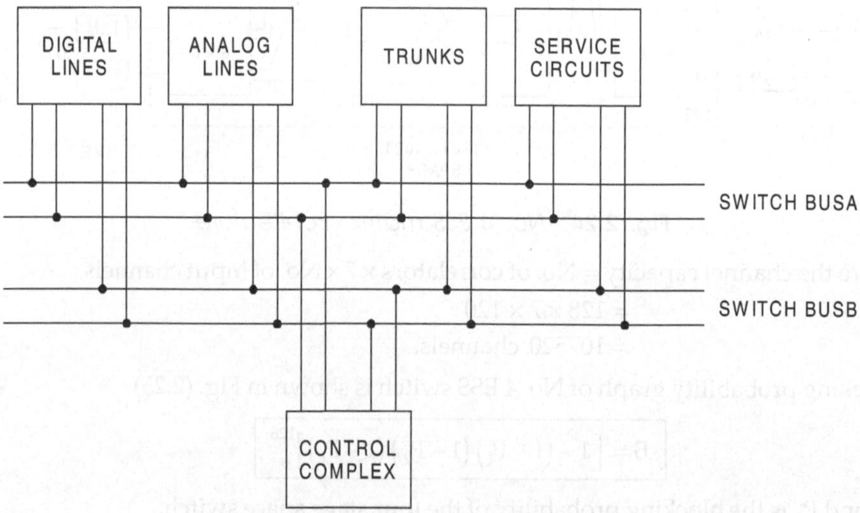

Fig. 2.26 *System 75 matrix architecture*

Configuration provide the facility of redundancy in case of any failure *i.e.*, all ports access both the buses. Thus, the blocking probability is zero. Basically congestion is responsible for blocking which can be controlled by congestion theory that deals with the probability that the offered traffic load exceeds some specific value.

2.21 DIGITAL CROSS CONNECT SYSTEM

In digital switching, in place of manual cross connect system, digital cross connect system (DCS) is used according to the requirements of network configurations. Basically DCS is a digital switching Matrix, follow the call connection request of the signalling information.

The Fig. (2.27) shows the manual cross connect system and Fig. (2.28) shows the digital cross connect system. The main purpose of cross connect system is to rearrange the transmission circuit and test the circuit in both directions. Generally, in two places manual cross connect system is used:

(*i*) **Switching offices:** The junction point between transmission facility and switching machines.

(*ii*) **Wire centers:** Between feeder cables and distribution facility.

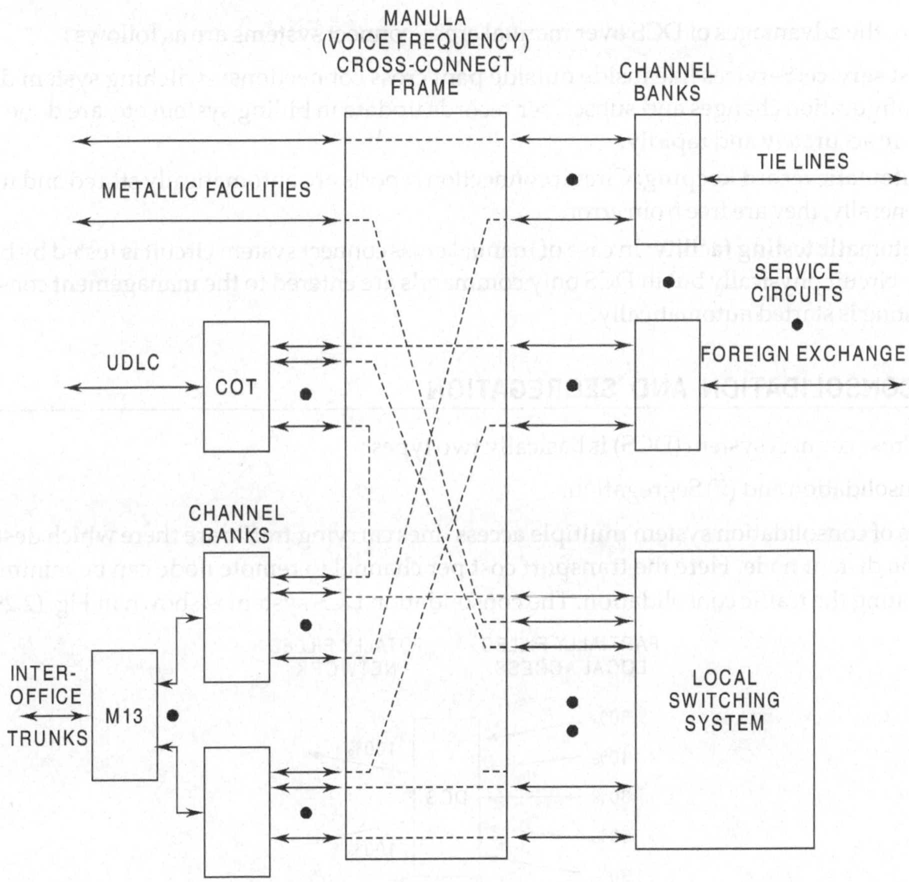

Fig. 2.27 *Manual cross-connect system*

In case of DCS cross connection is made by giving commands at the management console as shown in the Fig. (2.28).

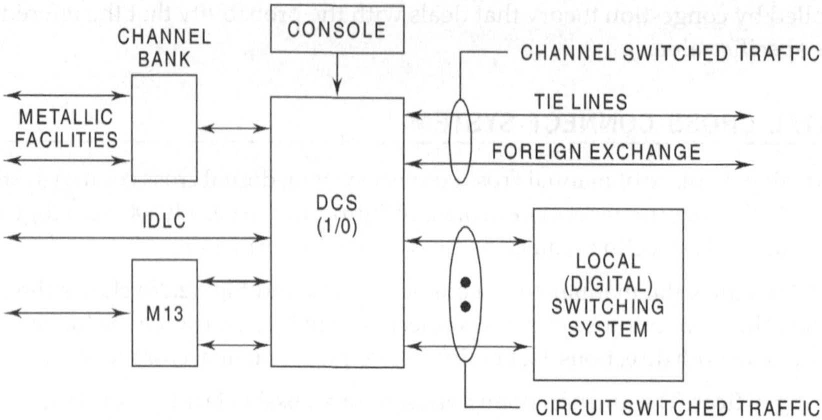

Fig. 2.28 *Electronic digital cross-connect system*

Some of the advantages of DCS over manual cross-connect systems are as follows :

1. **Fast service:** Services like inside outside pant cross connections, switching system database configuration changes and subscriber records update in billing system etc. are done by DCS more accurately and rapidly.
2. **Automatic record keeping:** Circuit connection reports are automatically stored and updated. Generally, they are free from error.
3. **Automatic testing facility:** In case of manual cross connect system circuit is tested by breaking the circuit physically but in DCS only commands are entered to the management console and testing is started automatically.

2.22 CONSOLIDATION AND SEGREGATION

Digital Cross connect system (DCS) is basically two types :

(*i*) Consolidation and (*ii*) Segregation.

In case of consolidation system multiple access lines carrying traffic are there which destined to a common distant node. Here the transport cost per channel to remote node can be minimized by consolidating the traffic consolidation. The consolidation DCS system is shown in Fig. (2.29).

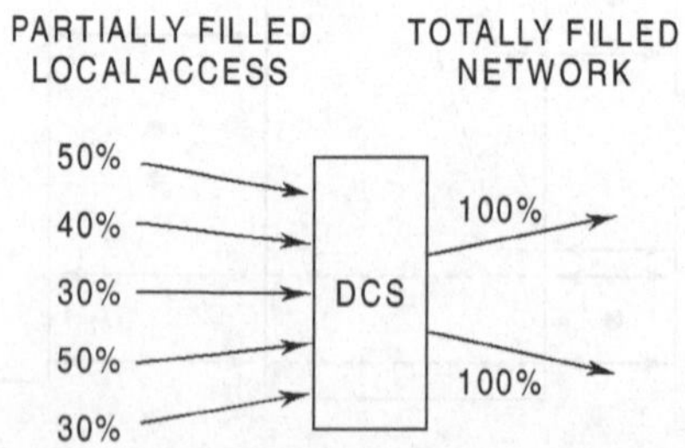

Fig. 2.29 *Consolidation of DCS system*

In case of segregation system, different types of traffic originate at a single location that allow to carry all types of traffic are destined to the DCS. Some of the example of such traffic are circuit switched channels, multidrop data circuit, FX (Foreign Exchance) circuits etc. Fig. (2.30) shows the segregation system.

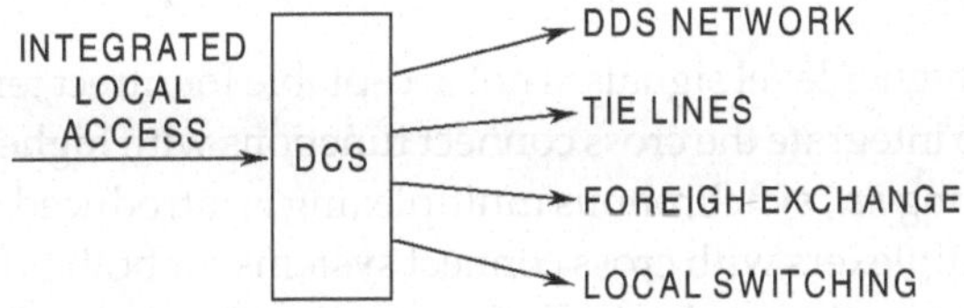

Fig. 2.30 *Segregation of DCS system*

2.23 DCS HIERARCHY

The digital cross connect system hierarchy is shown in Fig. (2.31). Basically, this provides rearrangement of lower signals that are not connected to the higher level signals. Here the system terminates DSO signals and interchanges DSO signals and this leads to the destination DCS 1/0. Similarly the higher level (DS3) signals, that are not cross connected to a lower level DCS may be unchannelized high speed (44.736 Mbps) signals or transit DS3s that terminate on DCS 3/1 or 3/o at a distant location. So, sometimes it is not necessary to provide cross-connect services at all levels of digital signal hierarchy for all the terminations. The main purpose of the DCS 3/3 is to provide network restoration or protection switching of the DS3 signals. Multiple levels of cross connect is also provided where in intermediate levels are also designated as DCS 3/1/0 for terminating DS3s and rearranging DS1 and DSO. Example of multiple level of DCS is the SONET system.

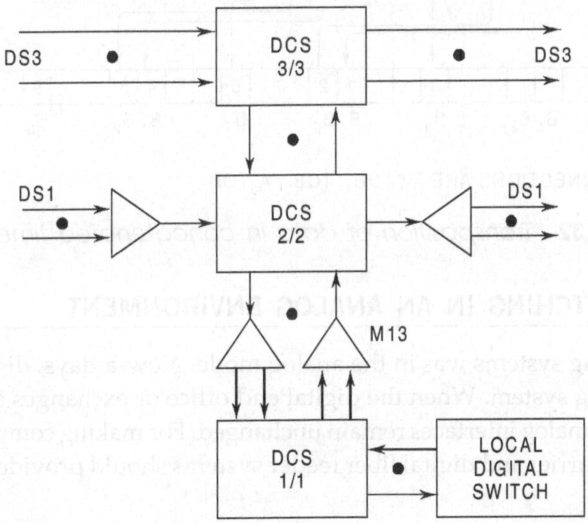

Fig. 2.31 *DCS hierarchy*

2.24 INTEGRATED CROSS CONNECT EQUIPMENT

Digital Cross Connect system have two distinct multiplexing equipment and two co-located switching systems— the DCS and Local Digital Switch.

When the DCS works in the public network then it use some traditional equipment but in case of private network it utilize equipment that provides integrated multiplexing and cross connect functions.

As the multiplexing of higher level signals is not acceptable for direct termination on a switching system, it is not possible to integrate the cross connect functions with higher level digital signals. So, a special type of multiplexing *i.e.,* synchronous multiplexing is introduced which allows integration of higher level digital multiplexers with cross connect systems for both public and private network equipment. Cross connect function are basically "pegged" or "nailed up" circuit switched connections.

But both the functions are different in the view of application aspect:

(*i*) DCS is uneconomical in large public network switches as it needs non-blocking at the DSO level.

(*ii*) DCS does not involve in processing of signaling bits and so interface cost is more.

(*iii*) Transparent cross connection of a 1.544 Mbps signal requires transport of the framing bit but DS1 interface equipment use framing in the 193rd bit.

(*iv*) The order of the data stream should be maintained and it is done by careful mapping of the individual 64-Kbps connections. This function does not provide by DSO circuit switching software. Transportation of data in concatenated time slot is shown in figure (2.32) below.

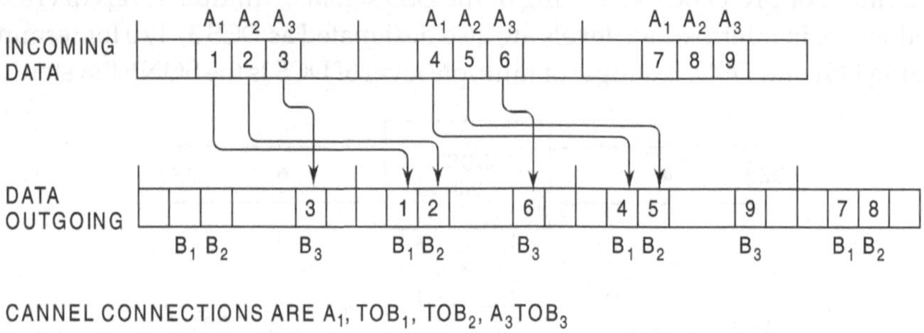

Fig. 2.32 *Transposition of data in concatenated time slots*

2.25 DIGITAL SWITCHING IN AN ANALOG ENVIRONMENT

All the earlier switching systems was in the analog mode. Now-a-days, digital switching system come in place of analog system. When the digital end office or exchanges are installed in analog environment, then the analog interfaces remain unchanged. For making compatible with the system the digital subscriber carrier and digital fiber feeder systems should provide analog interface with digital switch.

2.25.1 Zero-loss Switching

As we know in the digital system, loss is minimized means in case of switching the end-to-end transmission of voice, there will be no degration of voice quality or the quality of digitized voice will be very good. For this, the designing procedure of digital switching system should be well.

Zero-loss switching or digital transmission faces some significant problem when it works under analog environment of class 5 central offices. Analog end-office switches are bidirectional *i.e.,* two wire subscriber loop. However voice digitizations require separate paths for transmit and receive signals. Therefore class 5 digital end office switch need four wire connection.

When this four wire switches are introduced in the two wire analog environment, hybrids are used to separate the two way transmission. So, hybrids are required at each end of the internal digital sub network with the potential for echoes and singing. The figure (2.33) shows the two wire analog interfacing of four wire digital switch.

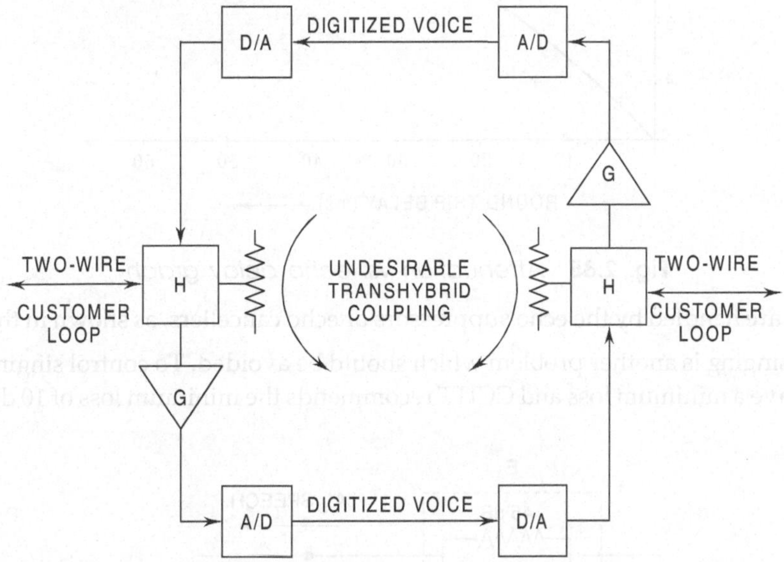

Fig. 2.33 *Four-wire digital switch with two-wire analog interfaces*

Echo shown in Fig. (2.34), is an important factor for creating disturbance to the speaker. Echo is nothing but the reflected component of the incoming signal. This eacho can be controlled by echo cancellation.

The transmitted signal is stored for a period of time equal to the round trip delay of the circuit. If the distance is short, the round trip delay experienced by the echo is small and create no disturbances. Small echoes can be attenuated by the help of attenuator. Fig. (2.35) shows the echo and attenuation versus echo delay graph.

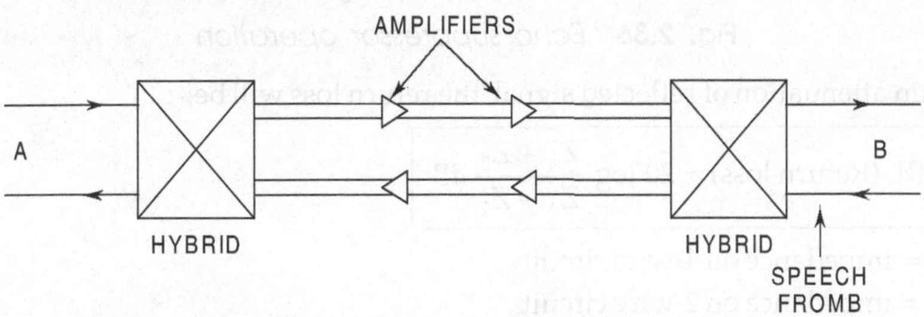

Fig. 2.34 *Echo as reflected signal*

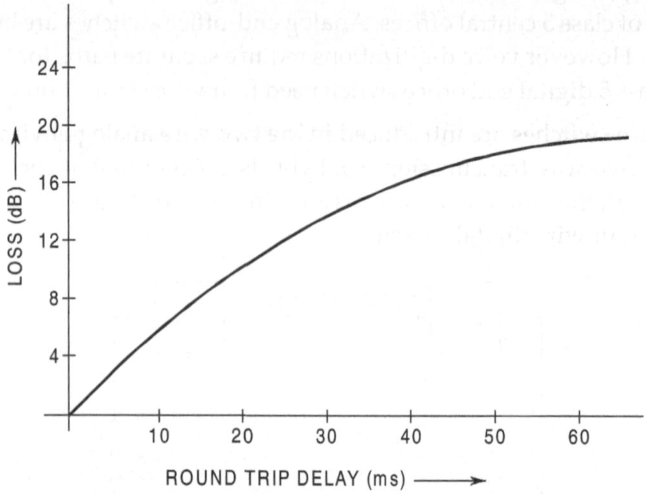

Fig. 2.35 *Attenuation vs. echo delay graph*

Long echos are handled by the echo suppressors or echo cancellers, as shown in the Fig. (2.36).

Moreover, singing is another problem which should be avoided. To control singing, the 4-wire circuit must have a minimum loss and CCITT recommends the minimum loss of 10 dB on national networks.

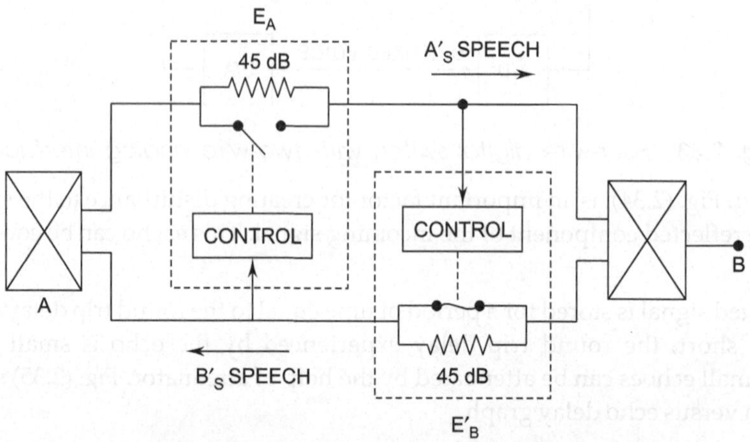

Fig. 2.36 *Echo suppressor operation*

For maximum attenuation of reflected signal, the return loss will be,

$$\text{RL (Return loss)} = 20 \log \frac{Z_4 + Z_2}{Z_4 - Z_2} \, dB \qquad \ldots(2.43)$$

where, Z_4 = impedance on 4-wire circuit

Z_2 = impedance on 2-wire circuit

Now, if the two networks are perfectly balanced then,

$$RL = 20 \log \frac{2Z_2}{0} \qquad [\text{As } Z_4 = Z_2]$$

So, $$\boxed{RL = \infty} \qquad ...(2.44)$$

This means that the return signal get infinite attenuation so, there will be no reflected signal. Therefore, we can say that end-to-end four wire digital circuit completely avoid echos and singing and make the circuit in such a way that if operates on zero loss mechanism.

2.25.2 BORSCHT Configuration

A set of functions is performed by the exchange to fulfil the requirement of signalling and voice transmission. Some functions are performed by digital network, some by analog network and some by both network. All these functions are summarized by the term BORSCHT, as follows

B	Battery feed
O	Over voltage protection
R	Ringing
S	Supervision
C	Coding
H	Hybrid
T	Test

Coding and hybrid functions are related to digital switching. In case of coding, analog to digital (A/D) and digital to analog conversion (D/A) is necessary and in case of hybrid, conversion of 2-wire line to 4-wire line is done. With the help of the transformer hybrid circuit is made.

Note. Reflection of voice signal at the listener's end is known as 'talker echo' and reflection of voice signal at the talker's end is known as 'listener echo'.

If multiple reflection occurs then oscillations are produced and the condition of the circuit become unstable. This condition is known as ringing.

SUMMARY

- Digital switching system also known as Electronic switching system-III generation and the switching process is by time division digital transmission.
- A switching system is called digital when the input to and output from the switching system can directly support digital signals.
- In India, C-DOT/MAX-XL type of digital switching system is used.
- In space division switching, the paths in the circuit are separated from each other.
- Time division switching involves the sharing of cross points for shorter period of time. Therefore get the alternative paths.
- Time division switching is equally applicable to either analog or digital switching.
- Combination of time and space switch is used in multistage switching systems.
- To calculate the blocking probability of multistage switching two methods after C. Y. Lee and C Jacobaeaus are used, referred as Lee graph method and Jacobaeaus method.
- A DCS is a digital switching matrix and its operation is to setting up relatively static connections between input and output signals/channels.

- No. 4 ESS toll switch can handle 120 channels TDM links formed by multiplexing five DS1 signals together.
- System 75 digital PBX uses full duplex connection and the operation speed is much faster.
- Blocking analysis of system 75 digital PBX can be analysed by congestion theory.
- Zero loss transmission provide some problems when digital switching is used in the analog transmission environment of a class 5 central office.
- Hybrids are introduced at the interface to the toll network where the four-wine transmission medium is used.
- End to end four wire circuit eliminates echoes and singing completely, allowing all connections to operate on a zero loss basis.
- Signalling and voice transmission on the subscriber lines requires that the exchange performs a set of functions (either in digital network or in analog network or in both), referred as BORSCHT.

REVIEW QUESTIONS

1. Define digital switching system. Give the classification of digital switching.
2. Explain the switching functions of digital switching.
3. Discuss the multiple stage switching.
4. What do you mean by analog time division switching ?
5. Differentiate between time division switching and space division switching.
6. Discuss briefly the two methods of blocking probability.
7. Write short notes on
   (*i*) path finding
   (*ii*) switch matrix control
   (*iii*) digital memory switch
8. What is TST ? Discuss its operation.
9. Compare STS and TST switch.
10. What do you mean by complexity ? Derive the implementation of complexity expression of TST switch.
11. Discuss the matrix architecture and blocking probability of No. 4 ESS toll switch.
12. Explain the terms
    (*i*) Consolidation and
    (*ii*) Segregation.
13. Discuss the system 75 digital PBX system.
14. What do you mean by digital switching in analog environment ? Explain.
15. Explain zero loss switching.
16. Briefly explain digital cross connect system.

MULTIPLE CHOICE QUESTIONS

1. If PCM binary samples are switched, then the switching is known as ___________
   (*a*) analog time division switching (*b*) digital time division switching
   (*c*) time division switching (*d*) None of the above.
2. No. 4 ESS toll switch is also known as ___________
   (*a*) TST switch (*b*) STS switch
   (*c*) TSTS switch (*d*) TSSSST switch

3. In ystem 75 digital PBX, the buses operate at ____________
   (*a*) 1.4 Mhz (*b*) 2.048 MHz
   (*c*) 5 MHz (*d*) 3.508 MHz.
4. The type of digital switching system used in India is ____________
   (*a*) AXE – 10 (*b*) GTD – SEAX
   (*c*) C – DOT / MAX-XL (*d*) EWSD.
5. The expression of blocking probability of STT switch is
   (*a*) $B = \left[1 - \left(1 - \frac{\rho}{\beta}\right)^2\right]^k$ (*b*) $B = \left(1 - \frac{\rho}{\beta}\right)^k$
   (*c*) $B = \left[1 - (1 - \rho T/k)^2\right]^k$ (*d*) $B = (1 - q^2)^k$
6. Hybrid circuit performs ________
   (*a*) 4 wire to 2 wire conversion (*b*) coding function
   (*c*) decoding function (*d*) 2 wire to 4 wire conversion
7. If there is no reflected signal, return loss is ________
   (*a*) 1.5 dB (*b*) 0.5 dB
   (*c*) Infinite (*d*) Zero.

ANSWERS

1. (*b*) 2. (*d*) 3. (*b*) 4. (*c*)
5. (*a*) 6. (*b*) 7. (*c*)

3

Telecom Traffic Engineering

3.1 INTRODUCTION

The switching system performs routing, flow control, security, failure monitoring, traffic handling, internetworking and network management etc. Within these functions, traffic handling is most important. As a designer or traffic engineer, one has to make a thorough study of the telephone traffic to decide how to handle all the subscribers smoothly/efficiently without fail with minimum cost as far as possible. Telephone traffic has similarities with road traffic as both are not uniform during all day (even interesting variation during reasonal and special occasion), so the traffic is unpredictable and random in nature.

Traffic engineering gives the ability of a telecom network to carry a given traffic at a particular loss probability. Traffic analysis based on advanced mathematical concept and research techniques. Now-a-days, advanced software tools are added to analyse the traffic pattern of an exchange.

3.2 NETWORK TRAFFIC LOAD AND PARAMETERS

In telecommunication system, network traffic load is expressed as distribution of traffic with respect to time. Time is normally 24 hours. The behaviour of the traffic as shown in fig. 3.1 which varies according to the specific period of time, reasons, festival, location of exchange, tourism area etc. Thus, the traffic is unpredictable and random in nature.

From the fig. 3.1 it is seen that during 0 and 6 hours when most of the population remain asleep, the traffic load is very little but around mid-forenoon, it is just opposite as it indicated busy office activities. Likewise afternoon and night time signify low traffic load. The characterize of the traffic load of some another day or another location may be different.

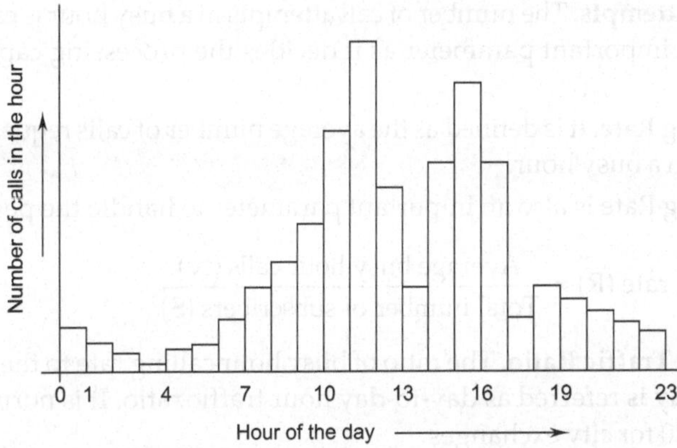

Fig. 3.1. *Telephone traffic pattern of a working day*

Now some basic terms or parameters related to network traffic are discussed below:

(*i*) **Busy Hour.** Continuous 1 hour period for which the traffic volume or the number of call attempts is greatest.

(*ii*) **Peak Busy Hour.** It is the busy hour each day varies from day to day, over a number of days.

(*iii*) **Time Consistent Busy Hour.** The 1 hour period starting at the same time each day for which the average traffic volume or the number of call attempts is the greatest over the days under consideration.

The busy hour varies from exchange to exchange depending on its location and its environment. Suppose an exchange located in a commercial area, the busy hour will be from 10-30 am to 11-30 am whereas the time differs if the exchange located in a residential area with amusement centers.

(*iv*) **Calling Rate.** Average number of requests for connection per unit time is known as calling rate. If '*n*' is the average number of calls to and from a terminal and time taken for that is T seconds then the calling rate will be,

$$\lambda = \frac{n}{T} \text{ calls/hour} \quad \text{...(3.1)}$$

(*v*) **Holding Time.** The average duration of occupancy of a traffic path by a call is termed as average holding time or service time "t_h". Therefore, the holding time consists of two parts—firstly, the time taken up by the exchange to set up the call and secondly, the time taken up by the actual conversation.

For voice traffic, the average holding time per call is in hours or seconds and for data traffic, average transmission per message is in seconds.

The reciprocal of average holding time is referred to as *service rate* (μ) in calls per hour *i.e.*,

$$\mu = \frac{1}{t_h} \text{ calls/hour} \quad \text{... (3.2)}$$

(*vi*) **Call Completion Rate.** The ratio of the number of successful calls to the number of call attempts is known as call completion rate (CCR). It is not necessary that all call requests or call attempts become successful. A CCR value of 0.75 is considered excellent but generally 0.70 is expected.

(*vii*) **Busy Hour Call Attempts.** The number of call attempts in a busy hour is called Busy Hour Call Attempts. It is an important parameter as it decides the processing capacity of a switching system.

(*viii*) **Busy Hour Calling Rate.** It is defined as the average number of calls requested or attempted by the subscriber—in a busy hour.

Busy Hour Calling Rate is also an important parameter to handle the peak hour traffic.

$$\text{Busy hour calling rate (R)} = \frac{\text{Average busy hour calls (N)}}{\text{Total number of subscribers (S)}} \quad \text{... (3.3)}$$

(*ix*) **Day-to-day Hour Traffic Ratio.** The ratio of busy hour calling rate to the average calling rate for a particular day is referred as day-to-day hour traffic ratio. It is normally 6 or 7 for rural areas and above 20 for city exchanges.

(*x*) **Traffic Intensity.** Traffic intensity is the ratio of the period for which the server is occupied to the total period of observation.

$$\text{Traffic intensity } (I_T) = \frac{\text{Period for which a server is occupied}}{\text{Total period of observation}} \quad \text{... (3.4)}$$

Generally, the duration of total period of observation is 1 hour. Traffic intensity is measured in two ways :

1. Erlangs (E) and 2. Cent call seconds (CCS)

The international unit of traffic is the Erlangs, named after A.K. Erlang, a Danish engineer. A server is said to have 1. Erlang of traffic if it is occupied for the entire period of observation. Assuming all servers are always busy, the maximum capacity of a group of servers in erlangs is equal to the total number of server. Erlang is expressed as 'E'.

Cent call seconds (CCS) is the call-time product. This is used as a measure of the amount of traffic expressed in units of 100 seconds. Sometimes call seconds (CS) and call minutes (CM) are also used as a measure of traffic intensity. The relation between Erlang and CCS is as follows—

$$\boxed{1E = 36\ CCS = 3600\ CS = 60\ CM} \quad \text{... (3.5)}$$

Note. CCS is used only for telephone circuit. For better measurement of traffic intensity 'erlang' is used as today's network support voice, data as well as many other services.

(*xi*) **Subscriber Traffic or Trunk Traffic.** The traffic intensity contributed by the subscriber or trunk is called subscriber traffic or trunk traffic. The network load or traffic intensity can be expressed as,

$$\boxed{A = R_a \cdot t_h} \quad \text{... (3.6)}$$

or

$$\boxed{A = \frac{n \cdot t_h}{T}} \quad \text{... (3.7)}$$

where, R_a = Average call arrival rate

n = Average noumber of call

t_h = Average holding time per call

T = Total period in seconds (generally T = 60)

Traffic not only generated by the subscriber but also busy servers create traffic. Sometimes load by the subscriber becomes more than the ca*pacity of the network. Then the traffic is called overloaded traffic. This situation can be handled by two different methods—

1. The overloaded traffic may be rejected instead of holding/keeping in a queue. So, this is the last case, where the users have to attempt once more *i.e.*, make a retry.
2. In the second case, a delay system is applied. In this case calls are not rejected but keeps in a queue until the network facilities become available.

Note. GOS or Grade of service and Blocking Probability are the e*xample of loss system. On the other hand, message network and packet network behaves as delay system.

Example 3.1. *In an exchange with 3000 subscribers, the total number of calls initiated in the busy hour is 12,000 and the call completion rate is 50%. Find the Busy Hour Calling Rate.*

Solution.

Given, Total number of subscriber = 3000

BHCA = 12,000

CCR = 50%

$$\text{Avg. Busy Hour Call} = \text{BHCA} \times \text{CCR}$$

$$= 12000 \times 0.50$$

$$= 6000 \text{ Calls.}$$

$$\text{Busy Hour Calling Rate} = \frac{\text{Average busy hour calls}}{\text{Total number of subscriber}}$$

$$= \frac{6000}{3000}$$

$$= 2$$

Example 3.2. *In an exchange with 5848 subscribers, the total number of calls originated in the busy hour is 6500. Calculate the calling rate.*

Solution. Total number of subscribers (S) = 5848

Total number of calls in the busy hour (N) = 6500/hour

$$\text{Calling rate (R)} = \frac{6500}{5848} \qquad \left[\because\ R = \frac{N}{S}\right]$$

$$= 1.11 \text{ calls/subscriber}$$

Example 3.3. *In a group of 15 servers, each is occupied for 60 minutes in an observation interval of two hours. Calculate the traffic carried by the group.*

Solution. $\text{Traffic carried per server} = \dfrac{\text{Occupied duration}}{\text{Total duration}}$

$$= \frac{60}{120} = 0.5 \text{ Erlang}$$

Total traffic carried by the group = 15×0.25

$= 7.5$ Erlang.

Example 3.4. *A group of 1000 subscribers which generate 500 calls during the busy hour. The average holding time is 2 minutes. Calculate the offered traffic in Erlangs, CCS and CM.*

Solution. Traffic intensity $A = \frac{n \cdot t_h}{T} = \frac{500 \times 2}{60}$ Erlangs

$= 17$ Erlangs.

Traffic intensity in CCS $= 17 \times 36 = 612$ CCS [$\because$ 1 E = 36 CCS]

Traffic intensity in CM $= 17 \times 60 = 1020$ CM [1E = 60 CM]

3.3 GRADE OF SERVICE AND BLOCKING PROBABILITY

The aim of an efficient switching exchange is to provide the best service *i.e.*, number call should fail thereby to provide as many lines as there are subscribers. But it is not economical. So, some calls have to be rejected and retried when the lines are used by other subscribers. Now, the grade of service is defined as the ratio of the unsuccessful or lost calls to the number of calls which has originated *i.e.*, sum of successful and unsuccessful calls. Thus, GOS is the ratio of lost traffic to offered traffic.

$$\text{GOS} = \frac{\text{Lost traffic}}{\text{Offered traffic}} \qquad \text{... (3.8)}$$

Now, offered traffic (A) = Average number of calls generated by the users × Average holding time per call.

Actual traffic carried by the network is called the carried traffic (Ao)

So, GOS is given by,

$$\boxed{\text{GOS} = \frac{A - A_o}{A}} \qquad \text{... (3.9)}$$

where,

A = offered traffic and

Ao = Carried traffic.

Thus a grade of service of 0.002 means 2 calls per 1000 or 1 in 500 has been lost. This is the recommended value of GOS. The smaller the value of GOS, the better is the service.

The switching exchange network is divided into following subsystems—

(*i*) an internal call *i.e.*, (subscriber to switching exchange)

(*ii*) an outgoing call to the trunk network *i.e.*, (switching exchange to trunk)

(*iii*) the trunk network *i.e.*, (trunk to trunk)

(*iv*) a terminating call *i.e.*, (switching exchange to subscriber)

Each subsystem has its corresponding GOS value. The overall GOS is the sum of all the subsystem GOS value.

The blocking probability (PB) is defined as the probability that all the servers (The elements of the hardware that make up a switching center and its associated facilities, are collectively referred to as servers) in a system are busy.

The congestion or blocking theory specify congestion in two ways. They are time congestion and call congestion. When all the servers in a group are busy, then the percentage of time is referred as time congestion. On the other hand, the call congestion is the proportion of calls attempting that do not find a free server. In connection with congestion, GOS is called call or demand congestion or loss probability and the blocking probability is called time congestion.

Note. Subscriber view point:

Grade of service (GOS) = Call congestion = Loss Probability

Network view point:

Blocking probability (PB) = Time Congestion

3.3.1 Difference between GOS and Blocking Probability

Grade of service (GOS)	*Blocking Probability (PB)*
(*i*) GOS is defined as the ratio of lost traffic to offered traffic	(*i*) Blocking probability is defined as the probability that all the servers in a system are busy.
(*ii*) GOS is related with subscribers view point.	(*ii*) Blocking probability is related with network view point.
(*iii*) GOS is arrived at by observing the number of rejected subscriber calls.	(*iii*) Blocking probability is arrived at by observing the busy servers in the switching system.
(*iv*) GOS is zero in case of equal number of subscriber and server.	(*iv*) Blocking probability is nonzero in case of equal number of subscriber and server.
(*v*) GOS is called call congestion.	(*v*) Blocking probability is called time congestion.

Example 3.5. *During a busy hour, 1500 calls were offered to a group of trunks and 15 calls were lost. The average call duration has 3 minutes. Determine (i) Traffic offered (ii) Traffic carried (iii) GOS and (iv) The total duration of period of congestion.*

Solution.

(*i*) Traffic offered, $A = \dfrac{1500 \times 3}{60} = 75 \text{ E}$

(*ii*) Traffic carried, $Ao = \dfrac{(1500 - 15) \times 3}{60} = 74.2 \text{ E}$

(*iii*) $GOS = \dfrac{75 - 74.2}{74.2} = 0.01$

(*iv*) Total duration $= 0.01 \times 3600 = 36$ seconds.

3.4 MODELLING SWITCHING SYSTEMS

In switching systems, traffic related all problems must be solved by through analysis. For this, a model, which gives mathematical expression of the behaviour of physical quantity under consideration, is necessary. A suitable model is choosen for handling a particular switching system and circumstance.

The behaviour of the switching system or network can be studied as random process. A random process or stochastic process can be defined as a process in which one or more quantities vary with time in such a manner that the instantaneous values of the quantities can not be determined properly but can be predicted under certain probability.

The quantities are called random variables. In a similar fashion, telephone traffic can be described as random process where the number of subscribers who generate the call and the number of busy servers are random variables. With certain probabilities, a prediction can be made for calculation of the number of active subscribers in a specific time interval. Fig. 3.2 shows the variations in the number of calls in a 30 minute interval or a random process.

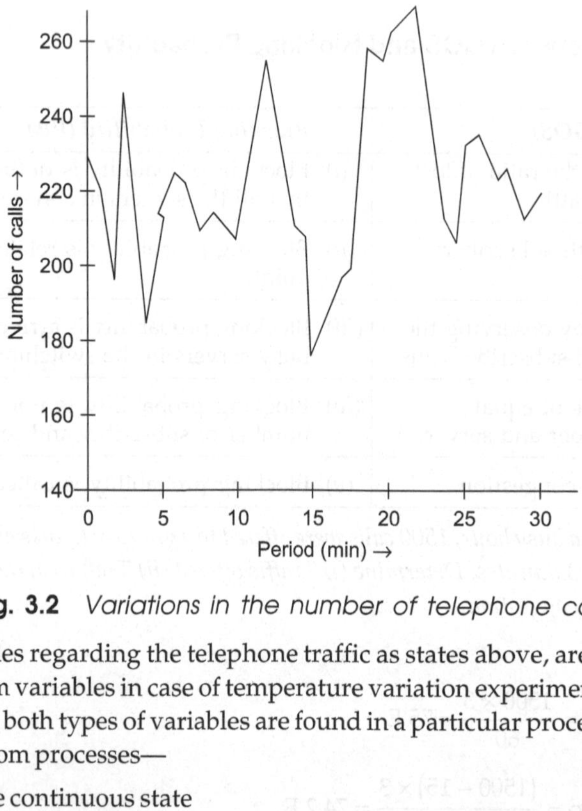

Fig. 3.2 *Variations in the number of telephone calls*

The random variables regarding the telephone traffic as states above, are discrete in nature. On the other hand, random variables in case of temperature variation experiment, can take on continuous values. Sometimes both types of variables are found in a particular process. Thus, there are four different types of random processes—

1. Continuous time continuous state
2. Continuous time discrete state
3. Discrete time continuous state
4. Discrete time discrete state

- The statistical properties obtained using continuous-time-continuous-state method are known as time statistical parameters.
- The statistical properties obtained using continuous-time-discrete-state method are referred as ensemble statistical parameters.
- A discrete state stochastic or random process is called a chain.

Fig. 3.3 illustrates the time and ensemble averages of stochastic process.

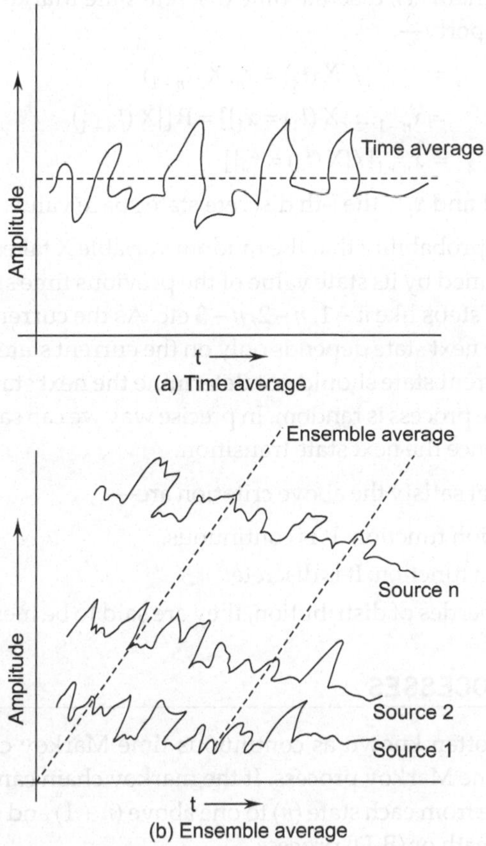

(a) Time average

(b) Ensemble average

Fig. 3.3 *Time and ensemble average of stochastic averages*

- The random processes with identical time and ensemble average are called ergodic process.
- For modelling, it is difficult for non-stationary process like telephone traffic. But during busy hour traffic may be considered as stationary or ergostic process.

Thus, it is seen that discrete state random process can be utilized in case of modelling as number of servers during busy hour or number of subscribers or number of calls are discrete random variables.

Note. The mathematical expression or description of the queuing system (subscribers attempting call waiting in a line or queue during busy hour) characteristic is referred as a queuing model.

3.5 MARKOV PROCESSES

A special property of random process was discovered by A. A. Markov in 1907. He observed a simple and highly useful form of dependency among the random variables forming a random process. So, Markov process is a random process, containing some special properties.

A discrete time Markov chain *i.e.,* discrete time discrete state Markov process is defined as one which has the following property—

$$P[\{X(t_{n+1}) = x_{n+1}\}/\{X(t_n) = x_n, X(t_{n-1}) = x_{n-1},\ldots,X(t_1) = x_1\}] = P[\{X(t_{n+1}) = x_{n+1}\}/\{X(t_n) = x_n\}] \quad \ldots (3.10)$$

where, $t_1 < t_2 \ldots\ldots < t_n < t_n + 1$ and x_i is the i-th discrete state space value.

Now, we can see that the probability that the random variable X takes on the value $x_n + 1$ at time step $(n + 1)$, which is determined by its state value of the previous time step n and is independent of its state values in the earlier steps like $n - 1, n - 2, n - 3$ etc. As the current state contains entire past history of the process so, the next state depends only on the current state. The total duration of time or the time period in the current state should not determine the next state. And the specification of time should be random as the process is random. In precise way we can say that the time taken for the current state does not influence the next state transition.

Distribution functions that satisfy the above criterion are—

(*i*) Exponential distribution function: It is continuous.

(*ii*) Geometric distribution function: It is discrete.

On the basis of these properties of distribution, they are said to be memoryless.

3.6 BIRTH-DEATH PROCESSES

Birth and death process is often known as continuous-time Markov chain as it a special case of discrete state continuous time Markov process. If the markov chain can occur only to the adjacent states *i.e.,* probability change from each state (n) to one above $(n + 1)$ and one below $(n - 1)$ it then the process is known as Birth-death or (B-D) process.

[The number of calls in progress is always between 0 to N so there will be $n + 1$ states]

In Telecommunication network, if number of busy servers represent the population and a call request means a birth and a call termination means a death then the network can be modelled as a B-D process. This concept or state transitions of birth-death process is shown in Fig. 3.4

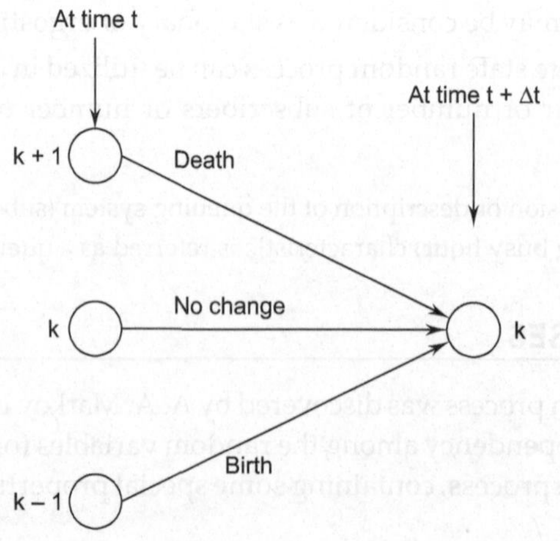

Fig. 3.4 *State transitions at a B-D process*

Let N (t) be a random variable which represents the total population at time t and assume N (t) is in state k at time t. The probability of state k is P (k) and probability of state (k + 1) is P (k + 1).

Thus, from Fig. 3.4 it is said that—

(*i*) The probability of state transition from state k to state $k+1$ in time duration Δt is $\lambda k\,\Delta t$ where λk is known as birth rate or call arrival rate in state k.

(*ii*) In the similar way, probability of state transition from state k to state $k-1$ in time duration Δt is $\mu k\,\Delta t$, where μk is called as death rate or call termination rate in state k.

(*iii*) The probability of no state change in time duration Δt is $[1-(\lambda k+\mu k)\,\Delta t]$.

(*iv*) The probability of state change from k other than $k+1$ or $k-1$ is zero.

Again, more specifically we can say that in the time interval Δt

(*a*) Probability of exactly one arrival is $\lambda.\Delta t$

(*b*) Probability of exactly one termination is $\mu.\Delta t$

(*c*) Probability of no arrival is $1-\lambda.\Delta t$

(*d*) Probability of no termination is $1-\mu.\Delta t$.

Now, the probability of finding the system in state k at time $t+\Delta t$ is given by,

$$P_k(t+\Delta t) = P_{k-1}(t)\,\lambda_{k-1}\,\Delta t + P_{k+1}(t)\,\mu_{k+1}\,\Delta t + (1-\lambda k\,\Delta t)(1-\mu k\,\Delta t)\,P_{k(t)} \qquad \text{... (3.11)}$$

From equation (3.11), it is seen that, the 1st term on the right hand side represents the probability of finding the system in $k-1$ state at time t and a birth occurring during time interval t, $t+\Delta t$. The second term represents the probability of binding the system in $k+1$ state at time t and a death occurring during time duration t, $t+\Delta t$. No arrival or no termination case is represented by the third term.

Now, ignoring the 2nd order term the equation (3.11) can be written as,

$$Pk(t+\Delta t) = P_{k-1}(t)\,\lambda_{k-1}\,\Delta t + P_{k+1}(t)\,\mu_{k+1}\,\Delta t - (\lambda_k+\mu_k)\,P_k(t)\,\Delta t + P_k(t) \qquad \text{... (3.12)}$$

$$\Rightarrow \quad \frac{P_k(t+\Delta t)-P_k(t)}{\Delta t} = P_{k-1}(t)\,\lambda_{k-1} + P_{k+1}(t)\,\mu_{k+1} - (\lambda_k+\mu_k)\,P_k(t) \qquad \text{... (3.13)}$$

In the limit $\Delta t \to 0$ we get,

$$\frac{d\,P_k(t)}{dt} = P_{k-1}(t)\,\lambda_{k-1} + P_{k+1}(t)\,\mu_{k+1} - (\lambda_k+\mu_k)\,P_k(t) \qquad \text{... (3.14)}$$

Equation (3.14) is the differential equation for all values of $k \le 1$, expressing the dynamic behaviour of a B-D process. Therefore, for $k=0$ the equn will be $\frac{d\,P_o(t)}{dt} = p_1(t)\,\mu_1 - \lambda_o\,P_o(t)$ ... (3.15)

[As, there can be no termination of a call *i.e.*, $\mu_o = 0$ and no state with -1]

Equation (3.15) give the rate of change of state probabilities. Under steady-state conditions, the state probabilities reach an equillibrium value and do not change with time *i.e.*, $P_k(t_1) = P_k(t_2) = P_k(t_i) = P_k$. Under this condition we get,

$$\frac{d\,P_k(t)}{dt} = 0 \qquad \text{... (3.16)}$$

and the B-D process becomes stationary. Thus, the steady-state equations about which we are concerned are—

$$P_{k-1}\lambda_{k-1} + P_{k+1}\mu_{k+1} - (\lambda_k + \mu_k) P_k = 0 \quad \text{for } k \geq 1 \qquad \text{... (3.17)}$$

$$P_1\mu_1 - \lambda_o P_o = 0 \quad \text{for } k = O \qquad \text{...(3.18)}$$

Therefore, with the help of these two equations (3.17) and (3.18), the system is modelled as a B-D process.

3.7 INCOMING TRAFFIC AND SERVICE TIME CHARACTERISATION

Traffic in the telecommunication network is the summation of individual traffic generated by the subscribers and it is characterized as a random process. When any subscriber generates a call it adds the number of the arriving calls. Once the number arrived, there is no way to reduce the number. So, it can be said that it is a special case of B-D process in which death rate is equal to zero. For this, we need a model that describes an originating process. Such a process is known as renewal process. In other words we can say, this process is a pure birth process that increases the population or traffic. The equations governing the dynamics of a renewal process can be easily arrived at from the B-D process equation [(3.17) and (3.18)] by $\mu_k = 0$. we have,

$$\frac{d\,P_k(t)}{dt} = P_{k-1}(t)\,\lambda_{k-1} - \lambda_k P_k(t) \text{ for } k \geq 1 \qquad \text{... (3.19)}$$

$$\frac{d\,P_o(t)}{dt} = -\lambda_o P_o(t) \ldots \text{for } k = 0 \qquad \text{... (3.20)}$$

In equation (3.17) and (3.20), the birth rate is dependent upon the state of the system. Let us assume a constant birth rate λ which is independent of the state of the system and then we get a Poisson process. The governing equations of a Poisson process are —

$$\frac{d\,P_k(t)}{dt} = \lambda P_{k-1}(t) - \lambda P_k(t) \text{ for } k \geq 1 \qquad \text{... (3.21)}$$

$$\frac{d\,P_o(t)}{dt} = -\lambda P_o(t) \text{ for } k = 0 \qquad \text{... (3.22)}$$

Now, we assume certain boundary conditions *i.e.*, at time $t = 0$, the system is in state zero or no births have taken place. So, we have,

$$P_k(0) = \begin{cases} 1 & \text{for} \quad k = 0 \\ 0 & \text{for} \quad k \neq 0 \end{cases}$$

Thus, equation (3.22) can be written as,

$$P_o(t) = e^{-\lambda t} \qquad \text{... (3.23)}$$

From equations (3.21) and (3.23) we get, for $k = 1$,

$$\frac{d\,P_1(t)}{dt} = -\lambda P_1(t) + \lambda e^{-\lambda t}$$

$$\Rightarrow \qquad P_1(t) = \lambda t e^{-\lambda t}$$

for $\qquad k = 2,$

$$P_2(t) = \frac{(\lambda t)^2 e^{-\lambda t}}{2!}$$

Thus, the general solution or equation is,

$$\boxed{P_k(t) = \frac{(\lambda t)^k e^{-\lambda t}}{k!}} \qquad \text{... (3.24)}$$

The equation represents the probability of *k* arrivals in the time interval *t* and equation (3.23) express the probability of zero arrival in a given time interval or probability distribution of inter arrival times *i.e.*, the time that elapses between two arrivals. So, in a Poisson Process, the inter arrival time is exponentially distributed as interstate transition times in case of Markov process.

Note. Steady state equations are not applicable to a renewal process. Suppose at time $t = 0$ we start our observation. As birth occurs say at time $t = t_1$ then it is totally impossible to find the system in zero state. Therefore, we concentrate here on the dynamics of the system.

Example 3.6. *A rural telephone exchange normally experiences six call originations per minute. What is the probability that exactly eight calls occur in an arbitrarily chosen interval of 60 seconds?*

Solution. $\lambda = \frac{1}{10}$ calls per second.

When $t = 60\text{s}, \lambda t = \frac{1}{10} \times 60 = 6$. So, the probability of exactly eight arrivals is given by,

$$P_8(60) = \frac{6^8 e^{-6}}{8!} = 0.1$$

3.8 POISSON ARRIVAL PROCESS

The equation (3.24) in section (3.7) *i.e.*, $P_k(t) = \frac{(\lambda t)^k e^{-\lambda t}}{k!}$ — is the Poisson arrival process equation which describe the probability of finding the system with k sources in the population at time *t*. It is essential to recognise that the assumption of an arrival rate independent of the state of the system in a Poisson Process as it may be infinite number of sources or constant number of sources. The busy source reduce the average arrival rate but in case of infinite population or large enough population, busy source does not affect that much.

3.9 HOLDING TIME OF CALLS

In conventional voice conversations the holding time is studied by an exponential distribution. Holding time may be constant in case of call processing and interoffice signalling. Let t_c be the constant holding time. Then the probability of finding the *k* active channels at any instant of time is the probability that there are *k* arrivals during the time interval tc immediately after the instant of observation that is given by,

$$P_k(t_c) = \frac{(\lambda t_c)^k e^{-\lambda t_c}}{k!} \qquad \text{... (3.25)}$$

Note. In B-D process the holding time or service time of calls have an exponential distribution or constant time distribution.

3.10 BLOCKING MODELS AND LOSS ESTIMATES

If the incoming calls finds all available lines busy, the call is said to be blocked. The blocked calls can be handled in two ways—one is loss system and another one is delay system.

In loss system, overflow traffic or blocked calls are simply refused or rejected or lost. The behaviour of loss system is studied by using blocking models.

In delay system, a blocked call remains in the system and waits in a queue for a free line. The behaviour of delay system is studied by queuing models.

In this section, we discuss about loss system, delay system will be described in the section 3.11.

The overflow traffic created in the loss system may be handled by three different ways. They are —

1. The rejected traffic by a particular resources may be handled by another resources in Telecommunication switching network.
2. After sometime the rejected traffic may return to the same resources.
3. The overflow traffic is actually serviced when the resources become available but it is virtually seen that the traffic is serviced by the resources. Here call is accepted and allowed to proceed with his information exchange process.

Considering these three cases, there are three models of loss systems —

1. Lost Calls Cleared (LCC)
2. Lost Calls Returned (LCR)
3. Lost Calls Held (LCH)

Lost Calls Cleared (LCC) are divided in two parts —

(*i*) Lost Calls Cleared System with infinite sources.

(*ii*) Lost Calls Cleared System with finite subscribers.

All the models are described in this section.

3.10.1 Lost Calls Cleared (LCC) System with Infinite Sources

In this case, the arrival rate is independent of the number of subscribers already busy and remains constant irrespective of the state of the system. The LCC model assumes that the subscribere hangs up the call whenever there is a busy or engaged tone and tries later. The reattempting call considered as a new call.

We can study the behaviour of trunk transmission system with LCC model. There are many trunk groups which are connected to different switching offices. So, when, direct trunk group between two switching offices are busy, then the overload traffic may be transferred to another switching offices via different trunk groups. The main purpose of the analysis is to estimate the blocking probability and grade of service. Using equation the offered traffic A for Poisson arrival process can be expressed as,

$$\boxed{A = \lambda . t_h} \quad \text{... (3.26)}$$

where, λ = average Poisson call arrival rate.

If all the servers in a system is not busy then the incoming traffic is accepted by the system otherwise it is simply rejected. This type of traffic is referred as Erlang traffic or pure chance traffic of type 1 and the arrival rate is referred as effective arrival rate. We also denote mean effective arrival rate as R_o and the effective arrival rate in state i as R_i. So, in this case, we have,

$$R_i = \lambda \text{ for } 0 \le i < S, R_s = 0$$

where S is the number of servers in the system.

Now, $$R_o = \sum_{i=0}^{S-1} \lambda P_i$$

where,

P_i is the probability that the system is in state i. The system can be in any one of 0, 1, 2, ... S states. Therefore we have,

$$\boxed{P_o + P_1 + P_2 + ... + P_s = 1} \quad ... (3.27)$$

Therefore,

$$\boxed{R_o = \lambda (P_o + P_1 + P_2 + ... + P_{S-1}) = \lambda (1 - P_s)} \quad ... (3.28)$$

The mean traffic carried by the n/w is given by,

$$\boxed{A_o = R_o t_h = \lambda (1 - P_S)\, t_h} \quad ... (3.29)$$

Now, equation (3.26) can be written as,

$$A_o = A (1 - P_S)$$

$$\Rightarrow \quad \boxed{P_s = \frac{A - A_o}{A}} \quad ... (3.30)$$

The blocking probability P_B is same as the probability that all the servers are busy *i.e.*, P_s. So, we can say from equation (3.29) and (3.30).

$$\boxed{GOS = P_B}$$

for LCC model where the traffic arrival is characterized by Poisson process.

Now for blocking probability, we have to study the steady state analysis of B-D process characterizing the LCC model. If a large number of servers are busy then the call termination rate will be higher. The termination rate is directly proportional to the number of busy servers as given by,

$$\boxed{\mu_k = k.\mu \quad \text{for} \quad 0 \le k \le S} \quad ... (3.31)$$

where μ = mean call termination rate = $1/t_h$

μ_k = call termination rate in state K

Using equation (3.17) and (3.18)

$$P_{k-1}\lambda + P_{k+1}\mu (k+1) - (\lambda + k\mu) P_k = 0 \quad ... (3.32)$$

Now, equation (3.26) can be written as,

$$P_{k+1} = \frac{A P_k + k P_k - AP_{k-1}}{K+1} \text{ for } k > 0 \quad ... (3.33)$$

$$P_1 = AP_o \qquad \text{for } k = 0 \qquad \text{... (3.34)}$$

For K = 1, we have,

$$P_2 = \frac{AP_1 + P_1 - AP_o}{2}$$

Substituting for P_1 from equation (3.34)

$$P_2 = \frac{A^2 P_o}{2}$$

Similarly for K = 2

$$P_3 = \frac{AP_2 + 2P_2 - AP_1}{3} = \frac{A^3 P_o}{3 \times 2} = \frac{A^3 P_o}{3!}$$

Therefore, general equation will be,

$$P_j = \frac{A_j P_o}{j!} \qquad \text{... (3.35)}$$

From equation (3.27) and (3.35)

$$P_o + AP_o + \ldots\ldots + \frac{A^s P_o}{S!} = 1 \qquad \text{... (3.36)}$$

$$\Rightarrow \qquad P_o = \frac{1}{1 + A + \frac{A^2}{2!} + \ldots\ldots + \frac{A^s}{S!}} \qquad \text{... (3.37)}$$

For K = S, substituting for P_o from equ^n (3.37) into equ^n (3.35)

we have,

$$\boxed{P_s = \frac{A^s/S!}{1 + A + \frac{A^2}{2!} + \ldots\ldots + \frac{A^s}{S!}}} \qquad \text{... (3.38)}$$

This equation (3.38) is known as Erlang B formula or loss formula. Here P_s represents the blocking probability of the busy servers and hence is the blocking probability P_B of the system. Therefore, in LCC model, GOS and blocking probability are same.

when S is large and A small then A/S is very small and hence equation (3.36) reduces to e^A. So, equation (3.36) will be,

$$P_j = \frac{A^j}{j!} e^{-A} \qquad \text{... (3.39)}$$

which is same as Poisson equation, where the traffic is Poisson. In the limits tending to infinity, we have

$$\lim_{S \to \infty} \frac{A^S e^{-A}}{S!} \to 0 \quad \text{or} \quad P_B \to 0 \qquad \text{... (3.40)}$$

Thus, the blocking probability tends to zero with Poisson traffic.

3.10.2 Lost Calls Cleared System with Finite Subscribers

In case of LCC model with infinite subscribers, call arrivals are independent of the number of active callers or subscribers as number of sources is much larger than the number of servers. But in case of LCC model with finite subscribers, the call arrival rate is dependent on the number of subscribers who are not occupied as the busy subscribers do not generate new calls. The traffic in this case is known as Engest traffic or pure chance traffic of type 2.

Let, λ_R = arrival rate per subscriber
K = number of busy subscribers.
S = number of server
N = total number of subscriber.

The offered traffic (arrival rate) when the system is in state k is given by,

$$R_k = (N - K)\,\lambda_R \qquad \text{for} \quad k \le 0 \le S$$

The mean offered traffic rate is given by,

$$R = \sum_{k=0}^{S} (N-k)\,\lambda_R\, P_k = N\lambda r \sum_{k=0}^{S} P_k - \lambda_R \sum_{k=0}^{S} kP_k$$

$$= \lambda_R \left(N - \sum_{K=0}^{S} kP_k \right)$$

$\left[\sum kP_k \text{ represents the average number of busy servers} \right]$

$$= \lambda_R (N - A_o) \qquad \text{... (3.41)}$$

The offered traffic is,

$$A = Rt_n = \lambda_R t_h (N - A_o) \qquad \text{... (3.42)}$$

when the system is in state S, the offered traffic rate is $(N-S)\,\lambda_R$, but all the arrivals are rejected. so, the lost traffic.

$$A - A_o = (N - S)\,\lambda_R\, P_R\, t_h \qquad \text{... (3.43)}$$

Therefore, the grade of service,

$$GOS = \frac{N-S}{N-A_o} P_R \qquad \text{... (3.44)}$$

So, we see that Engest traffic, the blocking probability and the GOS are not same or we can say that the time congestion and the call congestion values are different.

By analyzing the steady state characteristics of B-D process, we can calculate the blocking probability and grade of service. The expressions are as follows—

$$P_B = \frac{P^R \binom{N}{S}}{\sum_{k=0}^{R} P^k \binom{N}{K}} \qquad \text{... (3.45)}$$

and
$$\text{GOS} = \frac{P^R \binom{N-1}{S}}{\sum_{k=0}^{R} P^k \binom{N-1}{K}} \qquad \text{... (3.46)}$$

3.10.3 Lost Calls Returned System

LCR model based on the assumption that the blocked calls return to the system as a form of retries. So, that the offered traffic now comprises two components:

Offered traffic = new traffic + retry traffic

The following assumptions are made to analyse the LCR model with regard to the nature of returning calls.

1. Retries calls get the service even if multiple retries are required.
2. No new call is generated when a blocked call is being retried.
3. Time between call blocking and regeneration is random and statistically independent of each other.
4. Typical waiting time before a retry is longer than the average holding time. If retries are immediate, congestion may occur or the network operation becomes delay system.

Suppose a system with first attempt call arrival ratio of λ (say 100). If a percentage B (say I) of the calls blocked, B times λ retries (*i.e.*, 9 calls retries). Of these retries, a percentage B will be blocked again.

Thus, the infinite series, total arrival rate λ′ is given as,

$$\lambda' = \lambda + B\lambda + B^2\lambda + B^3\lambda + \ldots\ldots \qquad \text{... (3.47)}$$

$$\boxed{\lambda' = \frac{\lambda}{1-B}} \qquad \text{... (3.48)}$$

where B is the blocking probability.

Similarly we can say

$$\boxed{\lambda' = \frac{\lambda}{1-\text{GOS}}} \qquad \text{... (3.49)}$$

The effect of returning traffic is insignificant when operating at low blocking probabilities or grade of service.

Note. For a given value of GOS, the LCC model permits a larger offered load to the system than the LCR model.

For a given traffic load and the capacity of the system, the estimated blocking probability has lower value in LCC model than the LCR model.

3.10.4 Lost Calls Held System

In case of lost calls held system, the blocked calls are held in a queue and serviced when the necessary facilities become available. In this system, the total time is not dependent on the waiting time. The total time is determined by the average service time required. The sources need service continuously for a period of time whether or not the services are available. So, as soon as the service becomes free from other calls it handle the queued calls. Thus, this is a real time application by LCH model. Time Assigned Speech Interpolation (TASI) system is the example of LCH system.

If a number of calls blocked, a portion of it is lost until a server becomes free to service a call. This is shown in fig 3.5. Generally, telephone network does not operate in a last call held manner.

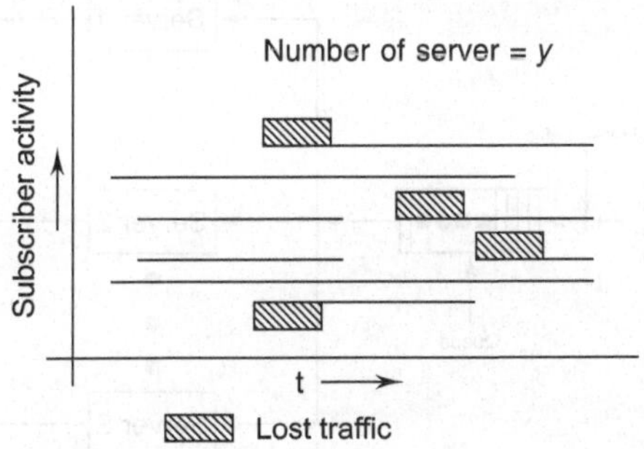

Fig. 3.5 *Lost traffic in LCH model*

LCH model easily studied to find the probability of the total number of calls in the system at a particular time. The number of active sources is identical to the number of call arrivals in that particular time. So, the distribution of the number in the system is the Poisson distribution, given as,

$$P(x) = \frac{\mu^x}{x!} e^{-\mu}$$

Example 3.7. *In a telephone system, there are 10 servers and 50 subscribers. On an average, there are 5 busy servers at any time. The probability of all the servers being busy is 0.2. Calculate the grade of service assuming—*

(a) *Erlang traffic and*

(b) *Engest traffic.*

Solution.

(a) As GOS is same probability of server PS in Erlang traffic. So, GOS = 0.2

(b) For Engest traffic,

$$GOS = \frac{N - S}{N - A_o} P_S$$

Carried traffic (A_o) = number of average busy server = 5

$$= \frac{50 - 10}{50 - 5} \times 0.2$$

$$= \frac{40}{45} \times 0.2$$

$$= 0.177$$

3.11 DELAY SYSTEM

In the absence of resources or servers, the arrived calls or messages are placed in a queue. Servicing is not taken up until the servers become available. Such system is known as delay system. It is also called as lost call delayed (LCD) system. For this system, queuing model is used as waiting line theory or queuing theory is applicable for this system. A wide variety of applications are there besides telecommunication. The examples are data processing (message switching, packet switching), aircraft landing, inventory control and various forms of services. An illustration of queuing system is shown in Fig. 3.6.

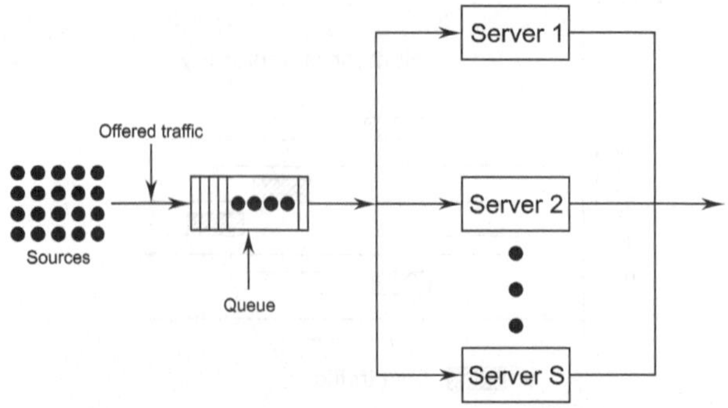

Fig. 3.6 *Elements of a queuing system*

By analyzing the delay system, we can determine the probability distribution of waiting times, which is a random variable and denoted by W. There are some factors on which waiting time W is dependent—

1. Number of sources
2. Number of servers
3. Intensity and probabilistic nature of the offered traffic.
4. Distribution of service times.
5. Service discipline of the queue.

There are two service time distributions. They are constant service times and exponential service times. In case of constant service time it is deterministic and for exponential it is random. The service of the queuing system include some basic factors—

(*i*) The selection of waiting calls in the queue may be determined by any one method like first-come-first served (FCFS) or first-in-first-out (FIFO), random selection and priority based selection method.

(*ii*) The second factor is the length of the queue. Delay or blocking probability based on the size of the queue in comparison with the number of effective sources.

A delay system with infinite queue capacity has a necessary condition for stable operation is,

$$\frac{\text{Mean arrival rate}}{\text{Mean service rate}} < 1$$

or,

$$\frac{\text{Offered traffic}}{\text{Number of servers}} < 1$$

We can model the Erlang delay system by B-D process where arrival (birth) and service (death) rates are as follows,

$$\lambda_k = \lambda \quad \text{for} \quad k = 0, 1, 2, \ldots..$$

$$\mu_k = k\mu \quad \text{for} \quad k = 1, 2, 3, \ldots.., S$$

$$\mu_k = S_\mu \text{ for } k > S$$

The stability condition for the system demands that $\lambda / S_\mu < 1$ and $A/R < 1$ ... (3.50)

Now we know that,

$$P_o + P_1 + P_2 + \ldots + P_s + P_{s+1} = 1 \quad \ldots (3.51)$$

[Modifying equation (3.27)]

So, we have,

$$P_o = 1 - \sum_{k=1}^{\infty} P_k \quad \ldots (3.52)$$

Now, substituting the B-D rates in equation (3.17) for $k = s$ and rearranging the terms we have,

$$S_\mu P_{S+1} = (\lambda + S\mu) P_S - \lambda P_{S-1}$$

$$\Rightarrow \quad S_\mu P_{s+1} = (\lambda + S_\mu) \frac{AS}{S!} P_o - \lambda \frac{A^{S-1}}{(S-1)!} P_o \quad \text{[Using the result of equ}^n \text{ (3.35)]}$$

$$\Rightarrow \quad S_\mu P_{S+1} = \frac{\lambda A^s}{S!} P_o \quad \text{[Using equ}^n \text{ (3.52)]}$$

$$\Rightarrow \quad P_{S+1} = \frac{\lambda}{\mu S} \frac{AS}{S!} P_o$$

$$\Rightarrow \quad P_{S+1} = \frac{A}{S} \frac{A \cdot S}{S!} P_o = \frac{A}{S} P_S \quad \ldots (3.53)$$

Similarly for $k = S + 1$ we get,

$$P_{S+2} = \left(\frac{A}{S}\right)^2 P_S$$

Thus,

general equation for $k > S$,

$$P_k = \left(\frac{A}{S}\right)^{k-S} P_S \quad \ldots (3.54)$$

Considering the inequality given in equation (3.50) we conclude that,

$$\frac{A}{S} + \left(\frac{A}{S}\right)^2 + \left(\frac{A}{S}\right)^3 + \ldots\ldots = \frac{A/S}{1-(A/S)} = \frac{A}{S-A} \quad \ldots (3.55)$$

From equation (3.51) and (3.55) we have,

$$\sum_{k=0}^{S} \frac{A^k}{k!} P_o + \frac{A}{S-A} \cdot \frac{A^S}{S!} P_o = 1$$

Therefore,

$$P_o = \frac{1}{\sum_{k=0}^{S} \frac{A^k}{k!} + \frac{A^S}{S!} \frac{A}{S-A}} \quad \ldots (3.56)$$

$$P_S = \frac{A^S}{S!} \frac{1}{\sum_{k=0}^{S} \frac{A^k}{k!} + \frac{A^S}{S!} \frac{A}{S-A}} \qquad \text{... (3.56a)}$$

Now, the probability of a message being delayed is nothing but the probability of finding the system in state S or above and is given by,

$$P(\text{delay} > 0) = \sum_{k=S}^{\infty} P_k \qquad \text{... (3.57)}$$

Thus, we have,

$$P(\text{delay} > 0) = \left(1 + \frac{A}{S-A}\right) P_S = \frac{S}{S-A} P_S \qquad \text{... (3.58)}$$

where, P_s is obtained by substituting the value of P_o from equation (3.56) in equation (3.38). Equation (3.58) is known as Erlang second formula, erlang delay formula or erlang C formula. If the term P_S in equation (3.38) as P_{SB} and P_S in equation (3.58) as P_{SC} then P_{SC} is related to P_{SB} as,

$$P_{SC} = \frac{P_{SB}(S-A)}{S - A(1 - P_{SB})} \qquad \text{... (3.59)}$$

Substituting for P_S in equation (3.58) we have,

$$P(\text{delay} > 0) = \frac{SP_{SB}}{S - A(1 - P_{SB})} \qquad \text{... (3.60)}$$

Where P_{SB} is the blocking probability given by Erlang B formula.

The exponential waiting time distribution is given by,

$$P(\text{delay} > t) = P(\text{delay} > 0) \exp[-(S-A)t/tn] \qquad \text{... (3.61)}$$

Integrating equation (3.61) over time the average waiting time may be determined as,

$$tq = \frac{P(\text{delay} > 0)\, t_h}{S-A} \qquad \text{... (3.62)}$$

Therefore, the waiting time of a call, already in the queue is expressed as,

$$\boxed{tq^1 = \frac{t_h}{S-A}} \qquad \text{... (3.63)}$$

3.11.1 Advantages of Delay System

Not a single subscriber want a delay in his/her call but it is better to have a service with a little delay rather than to have no service at all. Some advantages of delay system are as follows :

- It gives the better utilization of any trunk or a group of trunk.
- As there is no loss to the calls even at peak traffic hour, the traffic handling capability will be easy.
- Blocking probabilities are much less than those of loss system.

SUMMARY

- The telecommunication network design and analysis are done in the traffic engineering.
- Grade of service (GOS) is considered by observing the number of rejected subscriber calls whereas the blocking probability (P_B) is considered by observing the busy servers in the switching system.
- Markov process and birth-death process are the two important process for analysis of telecommunication network.
- The random processes which have identical time and ensemble averages are known as ergodic process.
- Renewal process is a pure birth process in which the death rate is equal to zero.
- The type of system by which a blocked call is simply refused and is lost is called loss system.
- There are three types of loss modelling system—
  LCC (Lost Call Cleared), LCH (Lost Call Hold) and LCR (Lost Call Return)
- The type of system by which a blocked call remains in the system and waits for a free line is called delay system.
- A delay system has infinite queue capacity and for its stable operation.

$$\frac{\text{Mean arrival rate}}{\text{Mean service time}} < 1$$

REVIEW QUESTIONS

1. What is the purpose of traffic engineering?
2. Define—Busy Hour, calling rate, holding time, traffic intensity.
3. Write short notes on—GOS and Blocking Probability.
4. Compare Grade of service and Blocking probability.
5. Comment on—modelling of traffic.
6. Explain briefly on Markov Process and Birth-Death Process.
7. What do you mean by incoming traffic and service time characterisation?
8. List the models of loss system.
9. Compare LCR and LCH system.
10. What is meant by delay? Compare it with loss system.
11. Give the expression of Erlang C formula.
12. Draw the queuing model of delay system.
13. Derive an expression that relates Erlang C formula.
14. Show state transitions at a B-D process.
15. Compare various types of blocking models system.

MULTIPLE CHOICE QUESTIONS

1. Grade of service or GOS of a network is given as

(*i*) $\text{GOS} = \frac{A_o - A}{A_o}$ (*ii*) $\text{GOS} = \frac{A - A_o}{A}$

(*iii*) $\text{GOS} = \frac{A}{A_o}$ (*iv*) $\text{GOS} = \frac{A_o}{A}$

2. Blocking probability is an example of
   (*i*) Loss system (*ii*) Delay system
   (*iii*) None.
3. The random processes which have identical time and ensemble averages are known as
   (*i*) ergodic processes (*ii*) ensemble statistical parameters
   (*iii*) time statistical process (*iv*) None.
4. Discrete time Markov process is geometrically distributed
   (*i*) Yes (*ii*) No.
5. In which type of models of loss system GOS = P_B?
   (*i*) LCC (*ii*) LCR
   (*iii*) LCH
6. For a given value of GOS, the LCC model permits a larger offered load to the system than the LCC model
   (*i*) True (*ii*) False.
7. TASI (Time Assigned Speech interpolation) system is an example of model
   (*i*) LCC (*ii*) LCH
   (*iii*) LCR
8. Message switching and packet switching are the example of which system
   (*i*) Delay System (*ii*) Lost System

ANSWERS

1. (*ii*)	2. (*i*)	3. (*i*)	4. (*i*)	5. (*i*)
6. (*i*)	7. (*ii*)	8. (*i*)		

4

Control of Switching Systems

4.1 INTRODUCTION

Earlier electronic switching system was based on the stores program control (SPC) system in which temporary memory was used for storing different information. At present, digital switching systems employ multiprocessor subsystem for best control. For that, softwares are necessary to implement the processing system for all the level of control structure.

In the digital switching system, call processing functions are performed by using interface controllers. Different processing steps and state transition concepts are discussed in this section.

Different software system used for switching control are also explained along with the stored program control concept. Microprocessor is mostly used for controlling in the digital switching system to reduce the cost of call processing.

4.2 CALL PROCESSING FUNCTIONS

Call processing in electronic exchange is similar to the common - control crossbar system except that most of the functions are performed by software or using interface controllers. The main system control process create and terminate some different software processes at the time of every call and handle the processing.

Some of the basic functions of call processing are—call identification, call routing, path setup between subscribers, digital translation, call status, billing etc. Sequence of operations are discussed in section (4.3). Generally, call processing takes place independent of the switching network.

4.3 SEQUENCE OF OPERATIONS

The basic steps involved in call processing is discussed as follows :

1. **Idle state:** In this first sequence the exchange is ready to accept any call request from the subscriber.
2. **Call request identification:** At the 'off hook' condition or when the subscriber lift the handset, there is a current (About 25 mamp) in the line which is called seized signal, indicate the exchange that there is a call request from the subscriber.
3. **Providing dial tone:** After receiving the seized signal, the exchange sends the dial tone to the calling subscriber to dial the number.
4. **Address analysis:** The exchange remove the dial tone after receiving the first digit and collect the whole number. Then address analysis is done like validity of the number, whether local, STD or ISD etc. for invalid number a recorded message is sent to the subscriber.
5. **Called line identification:** From the dialled number the exchange identify the called line or termination line.
6. **Status of called subscriber:** The called line may be busy or free or unavailable or out of service. In case of busy, number unobtainable or busy tone is sent to the calling subscriber. For other cases, recorded messages are sent to the calling subscriber.
7. **Ringing:** Whenever the exchange find the called subscriber free, power ringing is sent to the called subscriber and audible ringing to the calling subscriber.
8. **Path setup:** Once the called subscriber lifts handset, the ringing is removed and the exchange make the complete connection between the two subscribers.
9. **Supervision:** The exchange works as a supervisor during conversation to detect the end of the call for charging.
10. **Clear Signal:** Once the conversation is over, either subscriber may replace his/her handset. If the calling subscriber replaces his/her handset then a clear signal is sent to the exchange known as clear forward signal and for called subscriber the clear signal, known as clear backward signal sent to the exchange.

4.4 SIGNAL EXCHANGES

To study the behaviour of different control units in switching centre there are two diagrams—

1. Signal exchange diagram
2. State transition diagram

In this section state exchange diagram will be shown. In the next section *i.e.*, (4.5) we discuss about state transition diagram.

Generally, once the conversation is over, there are two types of difficulties arises—

1. **Called Subscriber held (CSH):** In this case, the called subscriber replace the handset but the caller does not. So, the caller can't make any other call or receive call from other subscriber.
2. **Permanent loop condition (PL):** This case is just opposite of CSH *i.e.*, caller replace the handset but called subscriber does not. So, the called subscriber can't get any other call. There will be a loop between called subscriber and exchange and for that any other calling subscriber get busy tone to this called subscriber. This type of situation is also called permanent glow condition in stronger switching system. This condition is overcomed by disconnecting the connection manually but in ESS system, a time out process is used.

If more than one trunk lines are used *i.e.*, many exchanges are used to setup a call between two subscriber, then the originating exchange to which the calling subscriber is connected sends the seize and then address to the terminating exchange to which called subscriber is connected. Remaining steps of signalling are same as local calls. The different call processing steps in case of local call is shown in figure (4.1). In SPC controlled exchanges, the inter exchange signallings are generated in originating exchange but processed at terminating exchange.

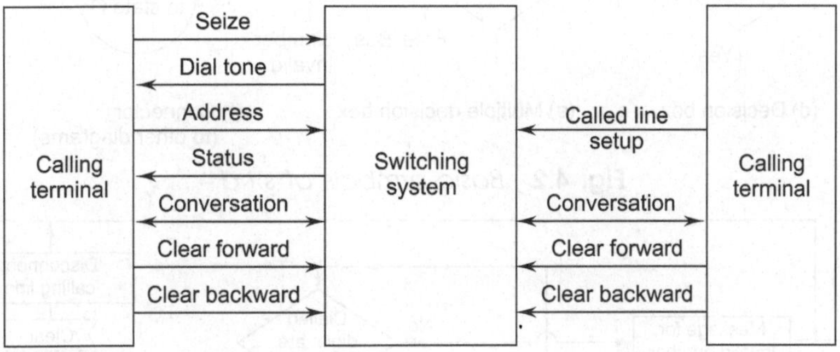

Fig. 4.1 *Signal Exchange diagram*

4.5 STATE-TRANSITION DIAGRAM

State-Transition diagram (s.t.d) is a design tool considering all possibilities of occurrance of every events. There are some basic symbols used in s.t.d. shown in fig. (4.2). An International standard for such diagram has been specified by CCITT; known as SDL (Specification and description language).

The basic symbols of s.t.d. are described briefly as follows :

State boxes: This boxes are labelled with state number and state description. More than one state occurs when the present state and a new task combine and results the next state.

Event boxes: The arrow symbol specify the event corresponds to the forward signal (signal from calling to called subscriber) or backward signal (signal from called to calling subscriber).

Action boxes: This boxes indicate the action taken on the event. The arrow specify whether the signal is sent forward or backward.

Decision boxes: In case of two division or multiple divisions of possibilities are represented by this diamond shaped decision box. In fig. (4.2) two division or multiple division decision boxes are shown.

Connectors: The connectors are used to connect one flow chart to another diagram.

The state transition diagram (s.t.d) of a local call is shown in the Fig. (4.3).

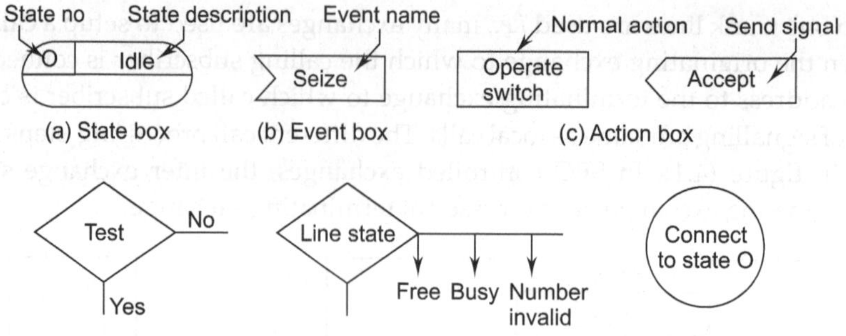

Fig. 4.2 *Basic symbols of s.t.d*

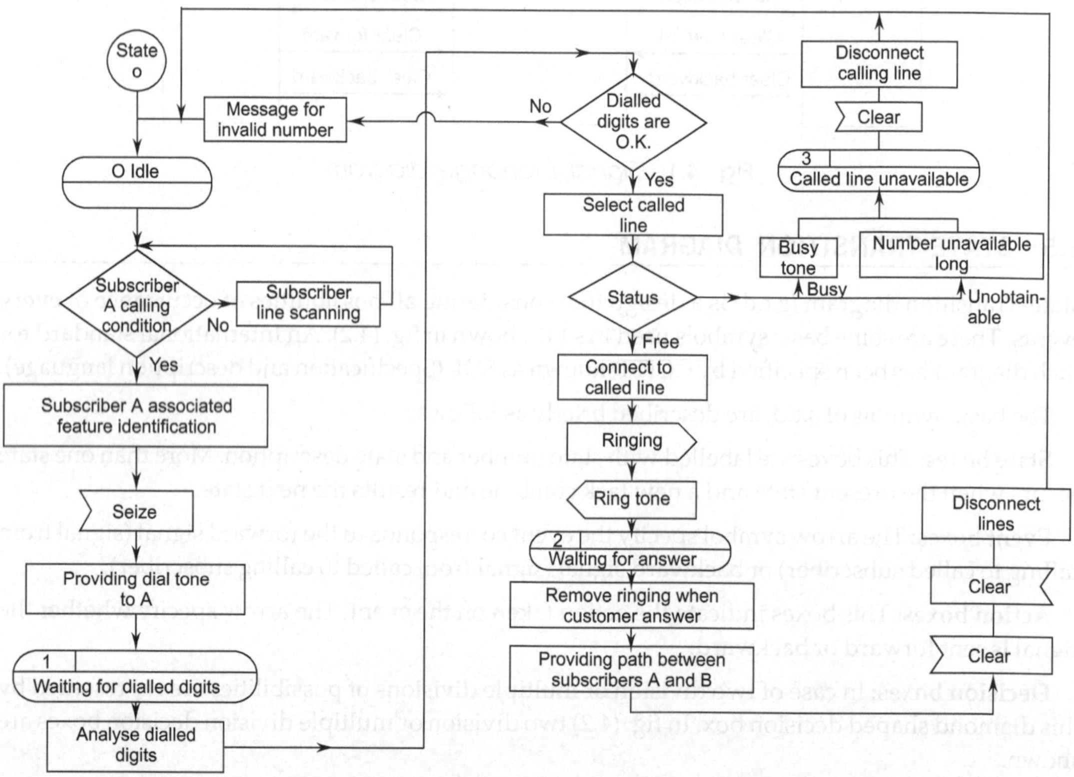

Fig. 4.3 *State transition diagram of a Local Call*

4.6 COMMON CONTROL

In the earlier time, there was a switching system mostly used, named as stronger switching system or step-by step switching system. In this switching system there was some disadvantages like

(*i*) for the entire call duration, the whole set of switches including contacts and relay mechanisms are busy and are not available for servicing other calls.

(*ii*) For a particular number, the subscriber location is also fixed. If the subscriber moves nearby location then he/she has to change the number.

(*iii*) Slow switching system

(*iv*) Responds only one subscriber at a time so limited facility.

All there disadvantages are overcomed by common control concept. In this method, there is no connection make until entire number is received. The number stored in a memory and the information conveyed by the number is translated by the logic circuitry into instruction to switch contact mechanisms which then contact the appropriate connections. After establishing the connection, the logic circuitry is free to service other calls. So, subscriber handling capacity is much more. Crossbar switching system which is discussed in chapter 1 (section 1.10), based on common control concept. Fig. (1.5) shows the common control switching system configuration.

Note. Those system in which the control subsystem is outside the switching network, are known as common control switching systems. Cross bar and electronic exchanges are common control system. It is also known as Indirect control or register control.

4.7 RELIABILITY, SECURITY AND AVAILABILITY

The term reliability is an important feature of any switching system. In case of common control and specially centralized control system the exchange depend on a small number of equipments. So that they must be designed for high reliability to obtain long mean time between failures (MTBF). In stored program control (SPC) system, the software must also be very reliable.

The another term security is also an important feature. Without security the existance of a system comes under doubt, therefore security should be provided in the switching system. Measured should provide security *i.e.*, to ensure correct operation even when faults are present in the system. The security measures that are used are as follows—

(*a*) Line circuits : none

(*b*) Switching network : none (or partial duplication)

(*c*) Common controls : 1 in n sparing

(*d*) Central processors : Replication.

As the terminating unit of a subscribers line contain very few components, so that it suffer less fault than the line itself. No additional measures are required to obtain the availability.

In the telecommunication system, if each incoming trunk has access to a sufficient no. of trunks on each route to give the required GOS is known as limited availability. The availability is defined as—

$$A = \frac{\text{MTBF}}{\text{MTBF} + \text{MTTR}}$$

where, MTBF = Mean time before failure

MTTR = Mean time to repair

4.8 STORED PROGRAM CONTROL (SPC)

In section (1.10), we discuss the crossbar switch where common control system was used. But with the development of modern electronic digital computer, it is realized that the registers and translators of the common control systems could be replaced by a single digital computer. The digital computer use the stored program concept, hence the name stored program control (SPC). The SPC concepts provide some significant advantages like call forwarding, call waiting, easy number changes, automated call tracing message unit etc.

4.8.1 Basic of SPC

In SPC, a program or a set of instructions to the computer is stored in its memory and the instructions are executed automatically one by one by the processor. with the help of these stored program, the exchange perform all the control functions. The control functions can be modified or extended by rewriting the program. Now all the public switching system use computers and software programming to control the switching of the calls. some features like common channel signalling (CCS), centralised maintenance and automatic fault diagnosis and human-machine interface are possible due to SPC to telephone switching. The figure (4.4) shows the basic control structure of SPC.

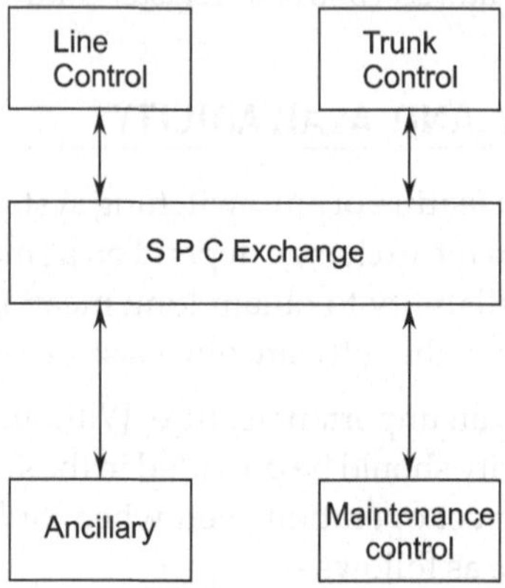

Fig. 4.4 *Basic control structure of SPC.*

4.8.2 Types of SPC

Basically, there are two types of SPC

(*i*) Centralised SPC

(*ii*) Distributed SPC.

Many exchanges use centralised SPC but with the development of powerful microprocessor and VLSI chips such as PLA, PLC etc. distributed SPC is mainly or popularly used.

4.9 CENTRALISED SPC

In centralised SPC, generally a single powerful processor is used and it must be capable of handling a number of calls depending on the load. Now-a-days, centralised SPC uses dual processor, each processor can access all the resources like printers, scanners etc and can control all the functions. The fig. 4.5 shows the typical centralised SPC configuration.

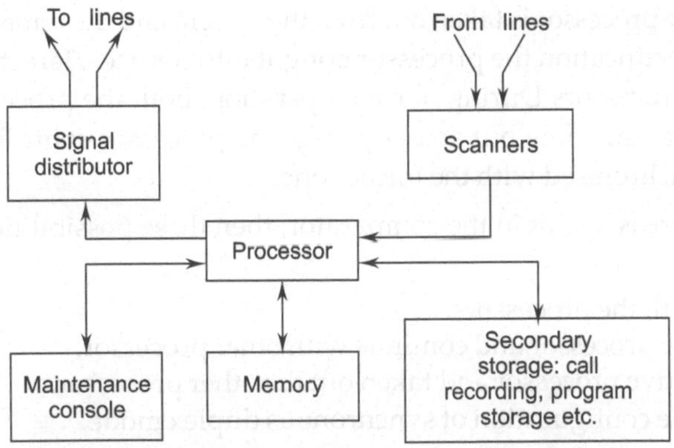

Fig. 4.5 *Configuration of centralised SPC*

A dual processor architecture may be designed to perform in three different modes—

1. Standby mode
2. Synchronous duplex mode
3. Load sharing mode.

All the three modes are described below:

4.9.1 Standby Mode

In this simplest configuration, any one of the processor is active and another one is standby until the first one (active processor) fail, both hardware and software basis. Here the prime work of the active processor is to copy the status of the system periodically and store in the secondary storage which is common to both the processors. When switchover occurs, the standby processor becomes online to load the most recent update of the system status from the storage and takes over the control *i.e.,* to determine which paths and subscribers and trunk lines are busy or free etc. In this case, the only calls which changed status in between last update and failure of the active processor are disturbed. The standby dual processor configuration is shown in Fig. 4.6.

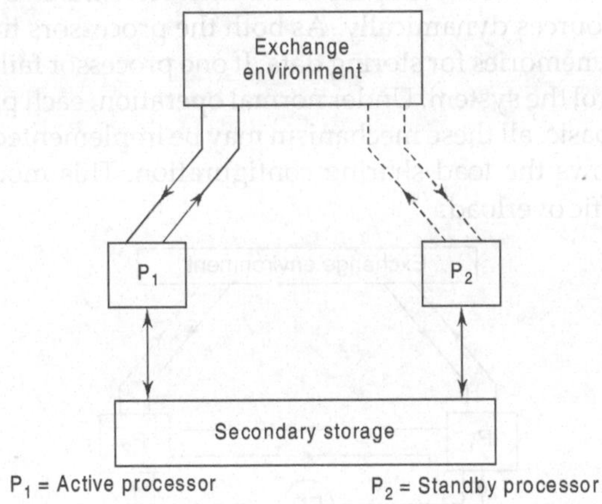

Fig. 4.6 *Dual Processor (Standby) configuration*

4.9.2 Synchronous Duplex Mode

In this mode, the two processors execute the same set of instructions and compare the results continuously in the comparator which is set in between the two processor. If, there is a mismatch in the result then the faulty processor is taken out from the system and the remaining processor control all the functions after rectification the processor bought into service. Here, two separate memories are used for both the processors During normal operation, both the processors receive the same information and execute the same but actually only one processor control the exchange and the another one remain synchronised with the former one.

It is possible that there is a fault in the comparator, then three possibilities exist to continue the system operation

1. Continue with both the processors.
2. Remove the active processor and continue with other processor.
3. Continue with active processor and taken out the other processor.

The fig. 4.7. shows the configuration of synchronous duplex mode.

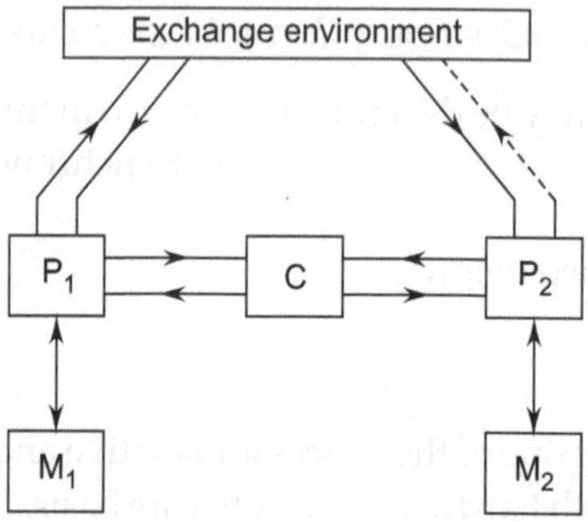

C = Comparator M = memory P = processor

Fig. 4.7 *Operation of Synchronous Duplex mode*

4.9.3 Load Sharing Mode

In the load sharing mode, in place of comparator, an exclusion device (ED) is used so that both the processors do not seek the same resource at the same time. Here the two processors are active and share the load and the resources dynamically. As both the processors handle the calls independently so there are separate memories for storing data. If one processor fails then the other processor takes all the load and control the system. Under normal operation, each processor handles one-half of the calls on a statistical basic. all these mechanism may be implemented in software or hardware or both. The Fig. (4.8) shows the load sharing configuration. This mode gives the much better performance in case of traffic overload.

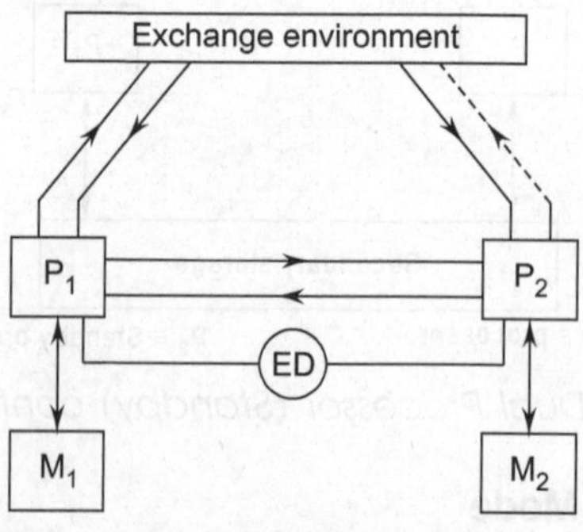

ED = exclusion device

Fig. 4.8 *Load sharing configuration*

Availability:

Single Processor :

$$\text{Availability (A)} = \frac{\text{MTBF}}{\text{MTBF} + \text{MTTR}} \quad ...(4.1)$$

where MTBF = Mean time between failures
MTTR = Mean time to repair

$$\therefore \quad \text{Unavailability (U)} = 1 - A = 1 - \frac{\text{MTBF}}{\text{MTBF} + \text{MTTR}}$$

$$\Rightarrow \quad U = \frac{\text{MTTR}}{\text{MTBF} + \text{MTTR}} \quad ...(4.2)$$

If MTBF >> MTTR, then

$$U = \frac{\text{MTTR}}{\text{MTBF}} \quad ...(4.3)$$

Dual Processor :

The MTBF of a dual processor is given by,

$$(\text{MTBF})_D = \frac{(\text{MTBF})^2}{2\,\text{MTTR}} \quad ...(4.4)$$

where $(\text{MTBF})_D$ = MTBF of dual processor.

A dual processor system is said to have failed only when both the processors fail and the total system is unavailable.

$$\text{Availability } (A_D) = \frac{(\text{MTBF})_D}{\text{MTTR} + (\text{MTBF})_D}$$

$$= \frac{\dfrac{(\text{MTBF})^2}{2\,\text{MTTR}}}{\text{MTTR} + \dfrac{(\text{MTBF})^2}{2\,\text{MTTR}}}$$

$$= \frac{(\text{MTBF})^2}{(\text{MTBF})^2 + 2\,(\text{MTTR})^2} \quad ...(4.5)$$

$$\text{Unavailability } (U_D) = 1 - A_D$$

$$= 1 - \frac{(\text{MTBF})^2}{(\text{MTBF})^2 + 2\,(\text{MTTR})^2}$$

$$= \frac{2\,(\text{MTTR})^2}{(\text{MTBF})^2 + 2\,(\text{MTTR})^2} \quad ...(4.6)$$

If MTBF >> MTTR

$$\therefore \quad U_D = \frac{2\,(\text{MTTR})^2}{(\text{MTBF})^2} \quad ...(4.7)$$

Example 4.1. *Calculate the unavailability for a single and dual processor systems for 5 years and 10 years if MTBF = 2000 hrs and MTTR = 4 hrs.*

Solution.

Given, MTBF = 2000 hrs.

MTTR = 4 hrs.

Unavailability of single processor,

$$U = \frac{MTTR}{MTBF} = \frac{4}{2000} = 2 \times 10^{-3}$$

for 5 years, $U = 24 \text{ yrs} \times 365 \text{ days} \times 5 \text{ years} \times 2 \times 10^{-3}$

for 10 years, $U = 24 \text{ hrs} \times 365 \text{ days} \times 10 \text{ years} \times 2 \times 10^{-3}$

$= 175.2$ hrs.

Unavailability of dual processor,

$$U = \frac{2\,(MTTR)^2}{(MTBF)^2} = \frac{2 \times 4^2}{(2000)^2} = 8 \times 10^{-6}$$

for 5 years, $U_D = 24 \times 365 \times 5 \times 8 \times 10^{-6}$

$= 0.35$ min

for 10 years, $U_D = 24 \times 365 \times 10 \times 8 \times 10^{-6}$

$= 0.7008$ hrs.

$= 4.2$ min

4.10 COMPARISON OF THREE MODES OF CENTRALISED SPC

Stand by mode	*Synchronous duplex mode*	*Load sharing mode*
(*i*) One processor is active and another one is stand by.	(*i*) Two processors are active and execute the same function.	(*i*) Two processors are active and execute the different function *i.e.*, share the load.
(*ii*) Common secondary storage is used.	(*ii*) Separate secondary storage is used.	(*ii*) Separate secondary storage is used.
(*iii*) In case of overload, performance is not good.	(*iii*) Performance is good in the overload traffic.	(*iii*) Much better performance in case of handling the overload traffic.
(*iv*) This mode has no capacity to handle large exchange.	(*iv*) Has good capacity to handle large exchange.	(*iv*) Increases the effective traffic capacity by 30 percent comparred to synchronous duplex mode.
(*v*) Purely centralised stored program control.	(*v*) Centralised stored program control system.	(*v*) One step forward to the distributed stored program control.

4.11 DISTRIBUTED SPC

The distributed SPC provide wider range of services with the help of many processors within the exchange itself. This configuration gives more availability and reliability. Entire exchange control functions may be configured either horizontally or vertically for distributed processing.

In the horizontal configuration, each processor performs one or some of the control function. Fig. (4.9) shows the typical horizontal configuration of distributed SPC. Here each functions are handled by each processor.

In the vertical configuration, the all control functions are devided into several blocks and each block is handled by a processor that perform all the functions related to that block. The different blocks are coupled together. for redundancy purpose, the processors may be duplicated to perform the functions in any three modes of dual processor. The Dual processor structure is also shown in fig. (4.10)

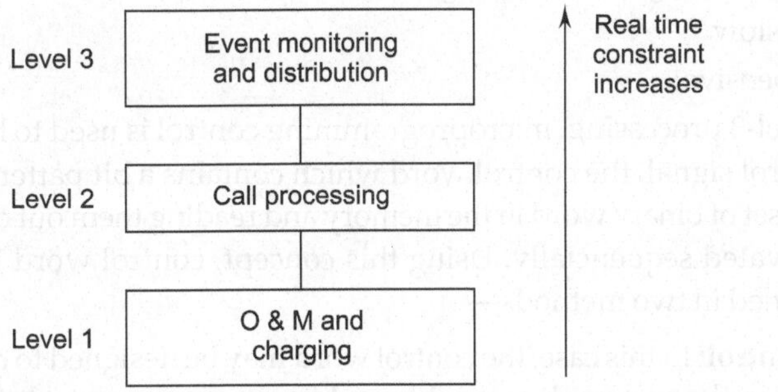

Fig. 4.9 *Horizontal Configuration of distributed SPC*

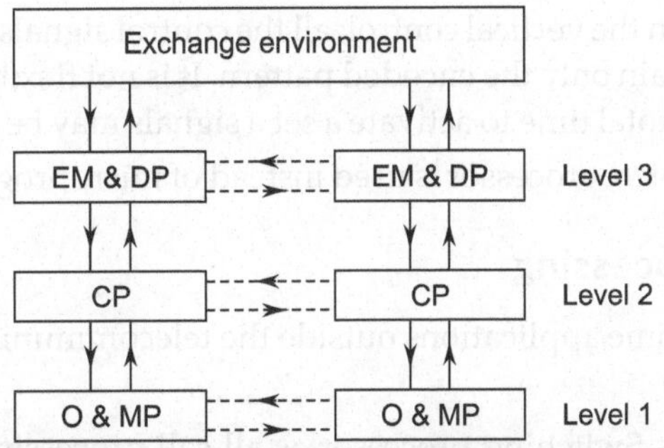

CP = call processor EM&DP = event monitoring and distribution processor O&MF = operation and maintenance processor

Fig. 4.10 *Dual Processor Structure*

4.11.1 Level-3 Processing

In the distributed SPC, each processor perform or specific task that's why they are designed to handle the functions efficiently. From fig. (4.9) it is seen that level-3 processor perform scanning, event monitoring and distribution functions as a specific task. In this level, it is necessary to sense or set or alter a set of binary conditions of a flip-flop or registers in a predefined sequence to control the functions. so, all the associated equipments are placed physically close to the switching network. To perform the operation smoothly, there are two processes used

1. Hard-wired control.
2. Microprogrammed control.

1. Hard-wired control:
   - It consist of logic circuits using logic elements, electronics.
   - Not flexible and can not be adapted to new requirements.

- For complex functions, implementation is not simple.
- Speed is high (Faster)
- Less expensive
- Difficult to maintain.

2. Microprogrammed control:
   - Universally accepted process control and easy to implement complex processing functions.
   - More flexible and easy to maintain.
   - Simply modifying the microprogram, it can handle different works.
   - Speed is slow
   - More expensive:

Generally for level-3 processing, microprogramming control is used to handle all the functions. To activate the control signal, the control word which contains a bit pattern of binary condition is altered. By storing a set of binary word in the memory and reading them out one after another, control signals may be activated sequencially. Using this concept, control word in a microprogrammed system can be designed in two methods—

1. **Horizontal control:** In this case, the control word may be designed to contain one bit per every control signal in the system. It is flexible and fast as many control signals may be activated simultaneously. Because of too large width of the control word it is more expensive.
2. **Vertical control:** In the vertical control, all the control signals may be binary encoded and the control word contain only the encoded pattern. It is not flexible as only one signal at a time is activated and the total time to activate a set a signals may be too large to realise.

In the latest design, microprocessor is used instead of microprogrammed unit as for low cost.

4.11.2 Level-2 Processing

In order to handle real-time applications outside the telecommunication field, level-2 processor is designed.

This is also known as Switching processor as all call processing are done here. Early general purpose computers are not suited for real-time applications. But fundamentally, switching processors are not different from general purpose digital computer. Certain specific characteristics are in the switching processor which enables the processor to process control and other industrial real time applications. One of the most important characteristic is that the processor is designed in such a way that within a short access time the memory is more tightly packet with data and single bit or half-byte manipulation instructions are used for switching applications.

The designing architecture of switching processor is such that it ensure over 99.9% availability, fault tolerance and security of operation. The I/O data transfer rate is not very high (in order of 100 kilobytes per second for large system). There are two technique for I/O data transfer

1. Program Control Data Transfer
2. Direct Memory Access (DMA).

Communication links are necessary if the exchange peripherals are located for away from the switching processor to connect them with the I/O controller.

The load on the switching processor is measured by the formula,

$$\boxed{t = a + bN} \qquad \text{...(4.8)}$$

where, t = occupancy

a = fixed overhead depending upon the exchange capacity and configuration.

b = average time to process one call

N = number of calls per unit time.

Generally the traffic load capacity is 40% higher than the normal load. In case of overload, the processor may be loaded to 95% of its capacity and traffic fluctuation can be absorbed.

Thus,
$$0.95 = a + 1.4b\,N_N \quad ...(4.9)$$

where, N_N = Nominal load in terms of number of calls per unit time.

$$\Rightarrow \quad N_N = \frac{0.95 - a}{1.4b} \quad ...(4.10)$$

The best way of measuring the processing time is to measure the actual processing time under this load as different instruction's processing time differ from one another.

4.11.3 Level-1 Processing

Operations and maintenance functions are handled by level-1 processing. Some important steps are there to maintain the functions, are as follows

1. Supervise the exchange hardware and software.
2. Add, modify or delete information in translation tables.
3. Change subscriber class of service.
4. Set a new line or trunk into operation
5. Supervise operation of the exchange
6. Watch the traffic.
7. Detect and locate faults and errors.
8. Run diagnostic and test programs.
9. Interact man-machine.

Generally, level-1 computer is a standard general purpose computer with large, volume of disk or tape storage due to the complex nature of the functions in comparison to level-2 and level-3 processing. It is less real-time constraints and less need of recurrent processing. Therefore, mainly shared among several exchanges located remotely as shown in fig. (4.11).

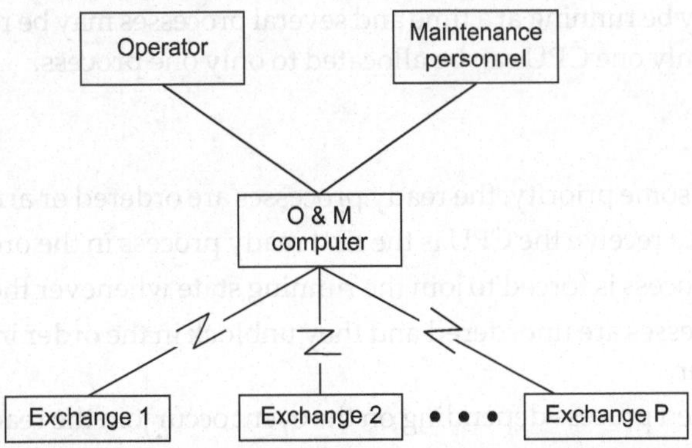

Fig. 4.11 *Remote operation and maintenance*

4.12 SPC SOFTWARE

Stored Program Control (SPC) software is same as general purpose computer. Basically, the software of SPC is two types

(*i*) System software (*ii*) Application software.

System software deals with processing system including language processor. And Application software deals with call processing, administration and maintenance. of the switching system.

4.13 SYSTEM SOFTWARE AND LANGUAGE PROCESSOR

In this context we will discuss the system software with call processing as call processing is an important part of switching system. It is a real time and even-oriented processing function. Each call is handled by separate process or task which after execution lasts for few tens to hundreds of milliseconds as waiting period of external events.

The most important feature of system software is that it works in powerful multiprogramming environment with thousands of processors simultaneously. A process in a multiprogramming environment may be in one of the following states

1. Running state (If CPU is allocated to it in current)
2. Ready state (If it could use one available CPU)
3. Blocked State (If it is waiting for any event to occur before it can process).

Different process states and transitions are shown in fig. (4.12)

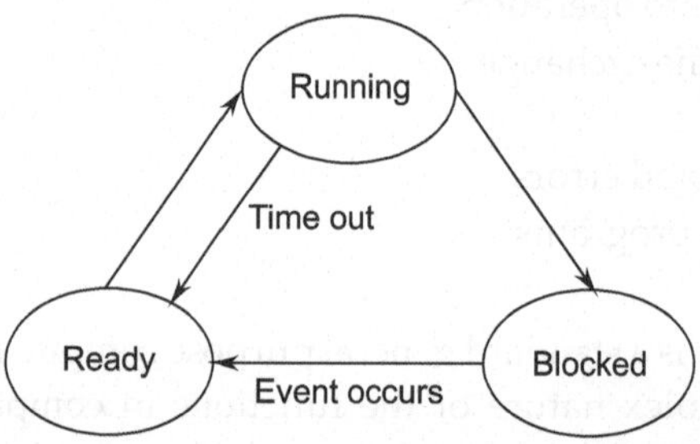

Fig. 4.12 *Process states and transitions*

Only one process may be running at a time and several processes may be ready and blocked state. At any instant of time only one CPU can be allocated to only one process.

Processing steps

(*i*) Depending upon some priority, the ready processes are ordered or arranged.

(*ii*) The next process to receive the CPU is the first ready process in the ordered list.

(*iii*) The first ready process is forced to join the running state whenever the timer runs out.

(*iv*) The blocked processes are unordered and they unblock in the order in which the events they are awaiting occur.

(*v*) Next, the unblocked process depending on the event occur join the ready state when the timer runs out.

According to the priority, the processes continues. Each process is represented in operating system by a process control block (PCB), a data structure, containing following information of the process—

1. Present State of the process and process priority.
2. CPU scheduling parameters and register save area to save the contents of the CPU register at the time of interrupt.
3. Memory allocated to the process.
4. Process account like the CPU time limits and usage, process number etc.
5. Status of the events and I/O resources associated with the process.

The CPU registers have program status word (PSW) that contains the next executable instruction, the status of interrupt (enabled or disabled) etc. The act of CPU switching between processes is illustrated in fig. (4.13). The process switching is also known as context switching which may occur on account of the currently running process becoming blocked or an event or interrupt that enable the high priority process. There are some priority level of an exchange in he decreasing order of priority, as follows—

- Fault alarms.
- Interprocessor communication, high speed peripherals (like disk etc).
- High speed clock interrupt for periodic tasks.
- Exchange peripherals.
- Low speed clock interrupt for periodic tasks.
- Call processing.
- Slow speed peripherals (terminals etc)
- O and M tasks.

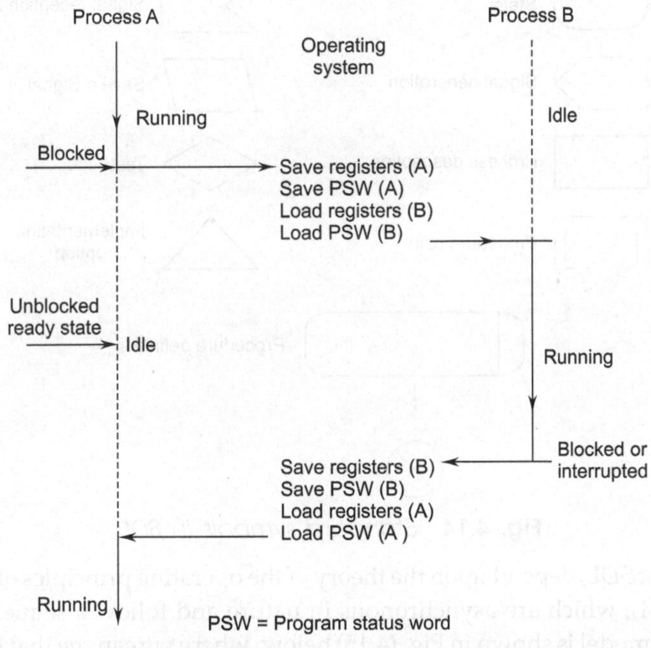

Fig. 4.13 *Process switching*

Now we will give our attention towards language processors *i.e.*, software. The design of software is recommended as top-down means proceeding from the general to the particular, with an increasing level of complexity and abstractions. In the designing process, the functional specifications of a system is described in natural language. Then to avoid the ambiguties of natural language, a formal language is used that also give the detailed specification, using diagrams.

The next step is the programming. This is a translation process *i.e.*, it convert all the formal specification into programming language. Programming may be divided into individual modules related to each other to work out easily. At last testing and verification are done to eliminate the error (usually bottom-up fashion) and to run the programme to handle the system.

According to the CCITT standards in the 'Z' series, some languages are used in the SPC system are as follows :

(*i*) Z.100 — Specification description language (SDL)
(*ii*) Z.200 — CCITT high level language (CHILL)
(*iii*) Z.3*xx* — Man-machine language (MML).

4.14 SDL (SPECIFICATION DESCRIPTION LANGUAGE)

With the help of the diagrams like flowchart, it briefly explain the functional specifications and the internal processes necessary to implement the specifications. Basically SDL is both a software development tool and a high level documentation technique. Based on the state transition diagram it represent sequential and conditional computations. SDL has a feature, called signal, to represent communication and synchronisation between concurrent processes. The standard symbols of SDL are shown in fig. (4.14), in which each diagram explains a single concurrent process.

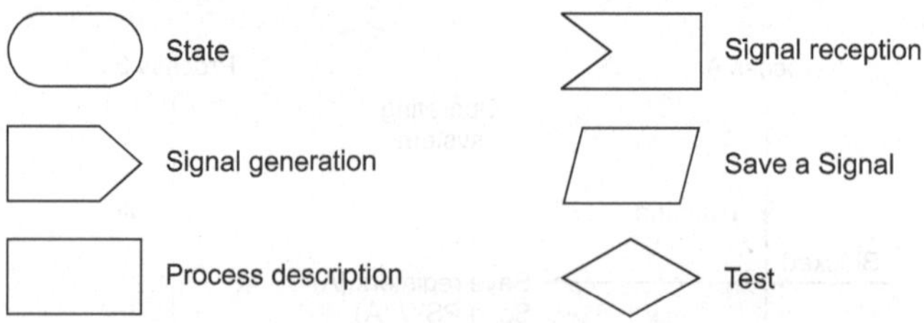

Fig. 4.14 *Standard symbols in SDL*

The basic design of SDL, depend upon the theory of the operating principles of synchronous finite state machines (FSM), which are asynchronous in nature and follow a sequential logic for their operation. The FSM model is shown in Fig. (4.15) below, where we can see that the status of output circuit not only depend on the input but also on the current state of the machine.

Switching systems belong to the class of FSM. As FSM is asychronous sequential operation, so that logic circuits and memory elements faces many problems due to transient variations. To overcome such problems clocked synchronous operation is necessary as shown in fig. (4.16).

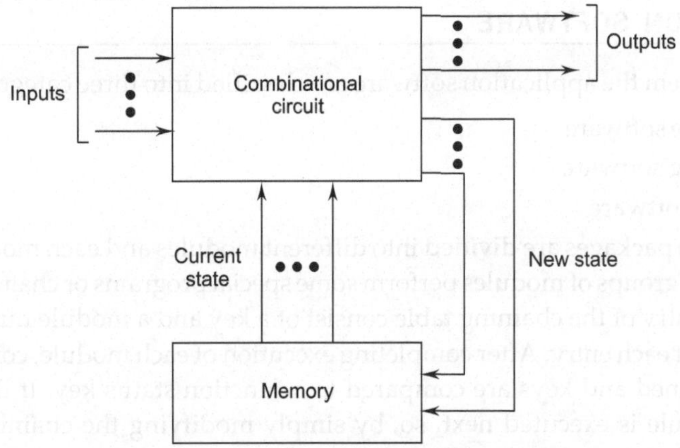

Fig. 4.15 *Finite state machine model*

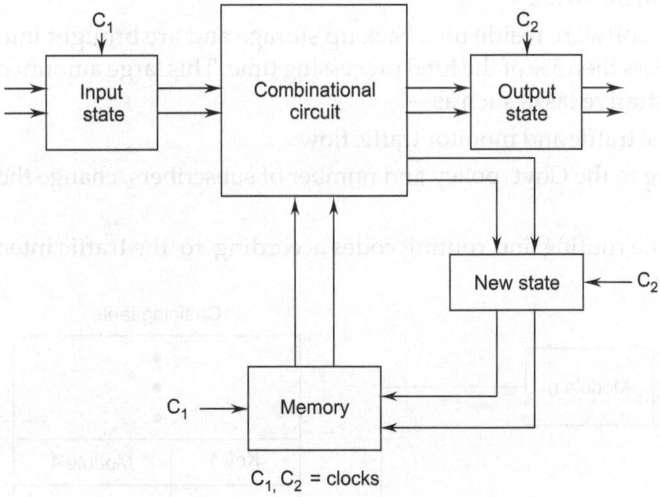

Fig. 4.16 *Synchronous FSM*

Advantages: Some advantages of SDL to both switching equipment manufacturer and operating and maintenance agency of the system as follows:

1. SDL provides unambiguous specifications and explanation for tendering and evaluation of offers
2. Oriented towards process interaction in a switching system.
3. It gives meaningful comparison of the capabilities of different SPC systems.
4. It is a good top-down design technique.
5. Various tools are available for storage, updating and logical verification of SDL specifications.
6. The SDL diagrams are easily transformed into excellent mathematical tool like Petri nets, for analysis and study of concurrent processes.

7. Translation of SDL specifications into CHILL code and vice-versa is available.
8. SDL is easy to learn, interpret and use.

4.15 APPLICATION SOFTWARE

In the switching system the application software are classified into three categories—

(*i*) Administrative software
(*ii*) Call processing software
(*iii*) Maintenance software

All these software packages are divided into different modules and each module perform a specific task. Sometimes groups of modules perform some special programs or chaining tables as shown in fig. (4.17). Each entry of the chaining table consist of a key and a module number and there is a pointer that point out each entry. After completing execution of each module, corresponding entries in the table are scanned and keys are compared to a function status key. If there is a match the corresponding module is executed next. so, by simply modifying the chaining table, there is a provision to add or delete modules to a functions.

(*i*) **Administration Software**

Generally this software reside on a back up storage and are brought into the main memory infrequently, less then 5% of the total processing time. This large amount of software perform some administrative tasks such as—

1. Record the traffic and monitor traffic flow.
2. According to the Govt. policy and number of subscribers, change the numbers and area codes.
3. Change the routing and routing codes according to the traffic intensity of a particular exchange.

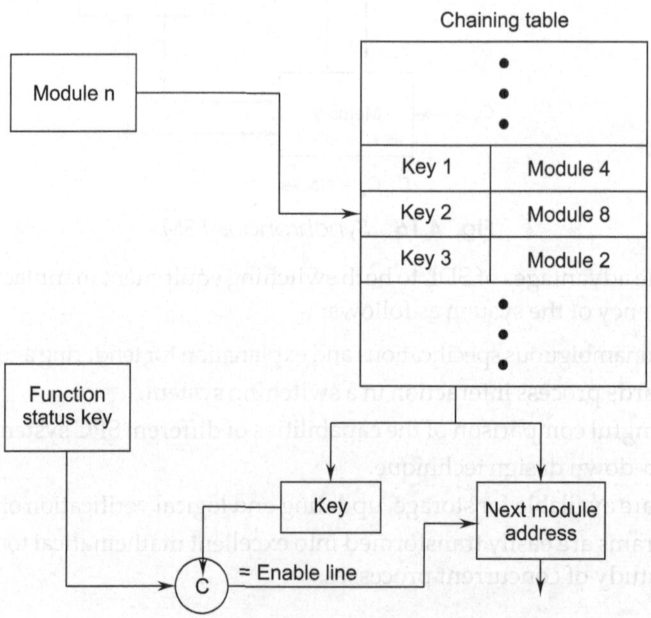

Fig. 4.17 *Module chaining through table*

4. Handle exchange management statistics.
5. Alarm processing.
6. Uncover terminal related problems.

Modern digital switching system perform some data base works to—

1. Record office information
2. Billing information.
3. Different parameters related to software and hardware.
4. System recovery parameters
5. Analysis or detailed information about the system.

Administrative software also known as Data base software.

(ii) **Call Processing Software**

Controlled by central processor and responsible for call processing. There are some system processes that run frequently and handle certain functions related to call processing like scans all the subscriber lines every few miliseconds looking for off hook conditions. When the condition is detected and reported to the main control process, other processes are activated to handle all the functions related to that particular calling line. The functions of looking up the subscriber line includes

1. Determine the type of dial *i.e.,* push button or DTMF push button
2. Allocate appropriate digit receiver.
3. Monitor the dialled numbers/digits.
4. Digit translation
5. Routing or set up switching n/w links.

The fig. (4.18) shows the main three levels of call processing program which varies with different digital switching systems.

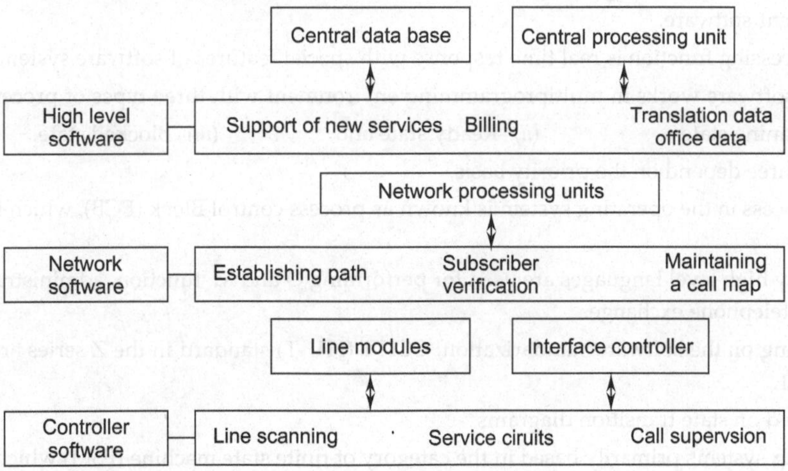

Fig. 4.18 *Call processing software levels.*

(iii) **Maintenance Software**

Maintenance software generally perform the preventive maintenance or diagnostic maintenance. Some of the main activities are

1. Monitor the proper functioning of the exchange equipment, trunks and subscriber lines.
2. In case of failure, best effort for system recovery.
3. Automatic line tests of various exchanges from one central location.
4. Apply effective maintenance stategies to reduce maintenance cost.

SUMMARY

- With the development of switching system *i.e.*, from manual to electronic, more services are provided to the subscribers and a particular function is performed with the help of specialized equipment that is connected in the common control environment.
- As the days passes, digital system is introduced in the switching system and provide the facility to control or perform different functions by the same hardware using different program. So, that stored program control takes the place of common control system.
- Centralised SPC is of three types—standby mode, synchronous duplex mode and load sharing mode. All the three modes are configured by dual processor architecture.
- Standby mode of operation is the simplest one.
- In distributed control, the control functions are shared by many processors within the exchange and the exchange control functions may be decomposed either horizontally or vertically.
- Level-3 processor controls the scanning, marking and distribution functions.
- Level-2 handles the call processing function.
- Level-1 control the operation and maintenance (O and M) functions for which it uses large disk or tape storage.
- SPC software may be categories as—system software and application software.
- System software architecture handles the general data processing system and application software controls the call processing function, Administrative function and maintenance function with the help of different software.
- Call processing function is real time response with special features of software system.
- System software works in multiprogramming environment with three types of processing steps— (*i*) Running state (*ii*) Ready state and (*iii*) Blocked state.
- All the three depend on the priority basis.
- Each process in the operating system is known as process control Block (PCB), which is a data structure.
- Generally high level languages are used for performing O and M function, administrative functions etc. in a telephone exchange.
- Depending on the software standarization, CCITT (ITU-T) standard in the Z series are SDL, CHILL and MML.
- SDL based on state transition diagrams.
- Switching systems primarily based in the category of finite state machine (FSM) which itself is asynchronous in nature and operates by following sequential logic.
- Application software holds 80% of the total volume of the software in a switching system.
- Application software packages of an Electronic switching system uses a modular organization.
- Administrative and maintenance software modules reside on a back-up storage *i.e.*, use virtual memory technique.

REVIEW QUESTIONS

1. What is state transition diagram? Draw the basic symbols of s.t.d.
2. Explain the concept of common control system.
3. What do you mean by stored program control? Mention different types of SPC.
4. Briefly explain centralized and distributed SPC and compare the two.
5. Describe the three level processing of distributed SPC.
6. Differentiate between system software and application software.
7. What is SDL? Explain.
8. What is language processor? Discuss.
9. Compare hardwired control and microprogrammed control.
10. Differentiate between three modes of centralized SPC.
11. Define the terms
    (*i*) Processor (*ii*) Task
    (*iii*) Chaining table (*iv*) PL
12. Derive the expression of availability of dual processor.
13. List the various standard symbols of SDL. Write some advantages of SDL.
14. Write the basic steps to process a call.
15. Briefly explain the working function of maintenance software.

MULTIPLE CHOICE QUESTIONS

1. SDL based on—
   (*i*) State Transition Diagram (*ii*) State Exchange Diagram
   (*iii*) Module Chaining table diagram (*iv*) None of the above.
2. The most simple mode of operation of dual processor architecture is—
   (*i*) Duplex mode (*ii*) Standby mode
   (*iii*) Load Sharing mode (*iv*) None
3. Practically which of the following architecture is used in the switching system—
   (*i*) Load sharing architecture (*ii*) Duplex architecture
   (*iii*) Standby architecture
4. Operation and maintenance (O and M) functions are handled by—
   (*i*) Level-1 processor (*ii*) Level-2 processor
   (*iii*) Level-3 processor (*iv*) None of the above
5. The unavailability of dual processor system is given by—
   (*i*) $\frac{\text{MTTR}}{\text{MTBF}}$ (*ii*) $\frac{(\text{MTTR})^2}{(\text{MTBF})^2}$ (*iii*) $\frac{2\,(\text{MTTR})^2}{(\text{MTBF})^2}$ (*iv*) None.

ANSWERS

1. (*i*) 2. (*ii*) 3. (*i*) 4. (*i*) 5. (*iii*)

5

Signalling

5.1 INTRODUCTION

Signalling systems are as essential as switching systems and transmission systems. With the suit able signalling system, a subscriber can be able to talk with or send data to someone in any where of the world almost instantly and the exchange is able to perform the instant switching requirements. The signalling system must be compatible with the transmission systems in order to reach the exchanges that they control. The design of signalling system is directly influenced by both switching and transmission requirements and evolution of signalling has come up with the developments in switching and transmission.

In this chapter, we will discuss, the various signalling techniques, popular signalling systems etc. which clear the basic concept of signalling.

5.2 CUSTOMER LINE SIGNALLING

All the subscriber in a telephone or switching network is connected to the nearest switching offices by a dedicated pair of wires. This pair of wires is known as subscriber loop. The Fig. (5.1) shows the hierarchy of cable for the subscriber loop. This subscriber loop is used for sending customer call signals from the exchange and clear or disconnect signal to the exchange. There is a minimum line current that the exchange can detect by a maximum line resistances. This limits the maximum length of the subscriber line and aria size served by the exchange.

The customer line signalling or subscriber loop signalling depends upon the type of telephone instrument/handset used. One is rotary dial telephone and another is push button or touch tone or key phone.

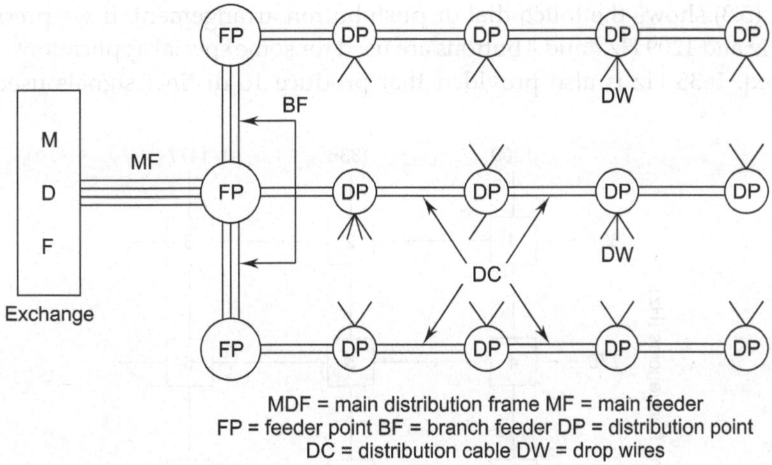

Fig. 5.1 *Cable hierarchy for subscriber loop.*

Depending on the mechanism used, the rotary dial telephone is of two types— (*i*) can type and (*ii*) trigger type. The operating principle is basically same for two types. Here a train of pulses is used to represent a digit in the subscriber number. The no. of pulse is equal to digit value except zero which is represented by 10 pulses. Two successive train of pulses are differentiated by a pause in between them known as interdigit gap. There pulses are generated by alternately making and breaking the loop circuit between the subscriber and the exchange. The Fig. (5.2 a, b) shows the Rotary dial telephone and pulse dialling system.

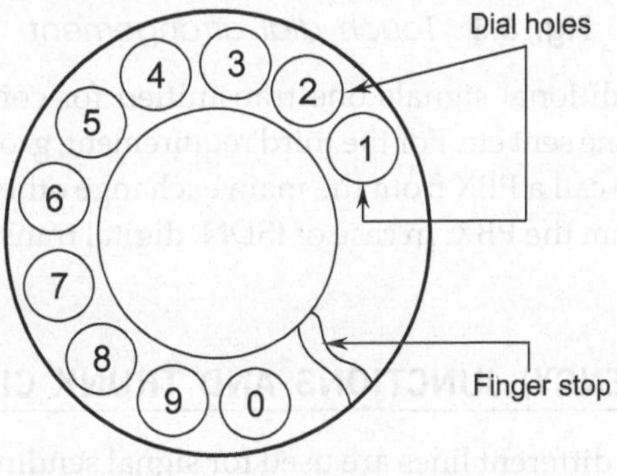

Fig. 5.2 *(a) Rotary dial telephone*

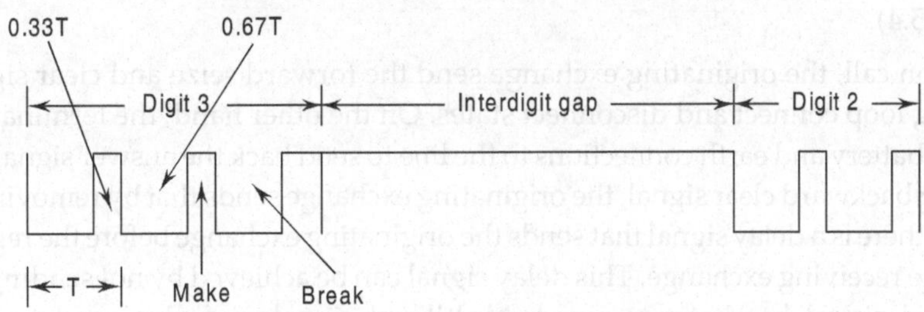

Fig. 5.2 *(b) Pulse dialling*

For faster dialling, touch tone dial telephone is used. Here 'touching' a button generates a 'tone' which is a combination of two frequencies, one from the lower band and the other from the upper band. The Fig. (5.3) shows the touch dial or push button arrangement. If we press no. 7 then it transmits 852 Hz and 1209 Hz. *and # buttons are used for some special applications. An additional upper band freq. 1633 Hz is also provided that produce 16 distinct signals used by millitary department.

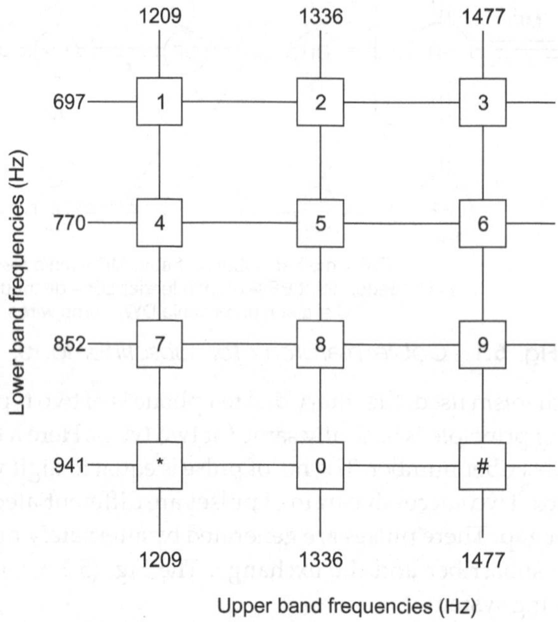

Fig. 5.3 *Touch dial arrangement*

In case of pay phone, additional signals one transmitted for coin collection and return and insertion of coin before dial tone sent etc. For the third requirement, ground-start-signal is applicable and this signal is also used to call a PBX from the main exchange otherwise subscriber line may be seized by an outgoing call from the PBX. In case of ISDN, digital transmission is used for customer access.

5.3 AF (AUDIO FREQUENCY) JUNCTIONS AND TRUNK CIRCUITS

For customer line signalling, different lines are used for signal sending and receiving between the caller and called subscriber and the exchange. A circuit for DC signalling over four wire AF line is shown in Fig. (5.4).

For a junction call, the originating exchange send the forward seize and clear signals and dial pulses by using loop connect and disconnect states. On the other hand, the terminating exchange interchange its battery and earth connections to the line to send back the answer signal. Again, at the time os sending backward clear signal, the originating exchange sends that by removing the polarity reversal. Now, there is a delay signal that sends the originating exchange before the register has been connected at the receiving exchange. This delay signal can be achieved by not sending digits until a proceed-to-send signal has been received. Multilevel signals are also used to send address information from automatic exchange to a manual exchange in more faster way than by 'loop connect/disconnect pulsing.

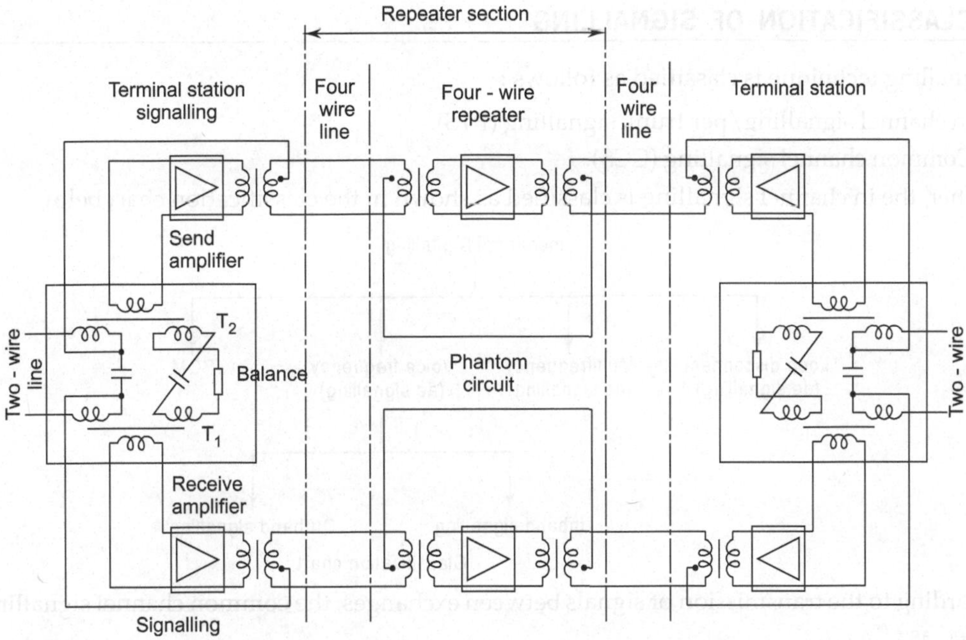

Fig. 5.4 *Circuit for dc signalling over Four-wire Audio Frequency (AF) line*

In case of long line, the capacitance of the subscriber line results a slow decay of the pulse waveform, so the output break from the receiving relay (use to make and brake contact) is shorter than that originally sent. This illustrate the Fig. (5.5) . Therefore distortion occurs. This pulse distortion can be reduced and the signalling range can be increased by using symmetrical waveform in case of Long Distance Direct Current (LDDC) signalling system as shown in Fig. (5.6). Here the pulses are received

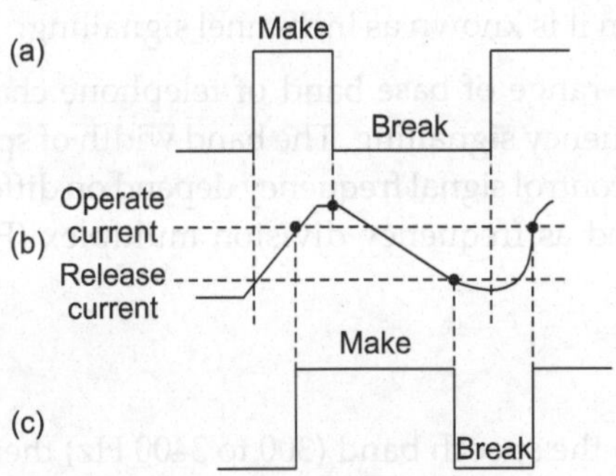

(a) Input pulse at sending end
(b) Arrival current Waveform
(c) Output pulse from receiving relay

Fig. 5.5 *Distortion of loop/ Disconnect pulses caused by distributed capacitance of a long line*

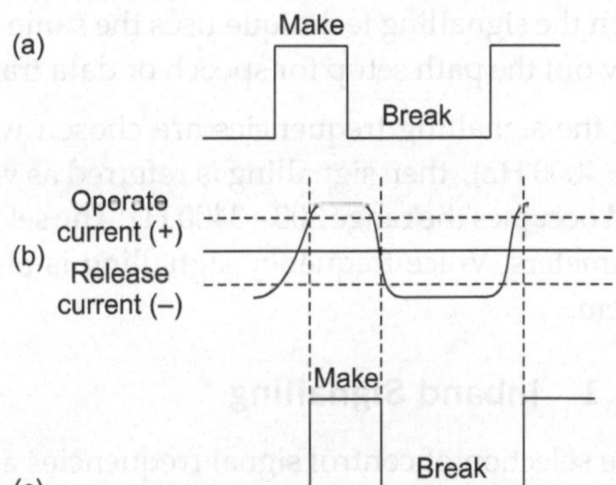

(a) Input pulse at sending end
(b) Arrival Current Waveform
(c) Output Pulse from receiving Polarized relay

Fig. 5.6 *Pulsing by current Reversals (Double current Working)*

by polarized relay. Most long distance circuits uses multiplex transmission systems with high capacity.

5.4 CLASSIFICATION OF SIGNALLING

The signalling technique is classified as follows :

(*i*) In channel signalling/per trunk signalling (PTS).

(*ii*) Common channel signalling (CCS).

Further, the in channel signalling is classified as shown in the classification chart below :

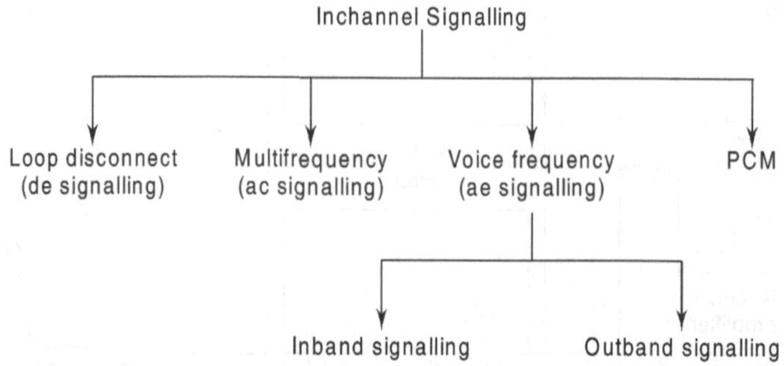

Classification chart

According to the transmission of signals between exchanges, the common channel signalling is classified as :

1. Associated signalling.
2. Non-associated signalling
3. Quare associated signalling.

5.5 INCHANNEL SIGNALLING

When the signalling technique uses the same channel to carry voice or data and control signals to carry out the path setup for speech or data transfer then it is known as inchannel signalling.

If the signalling frequencies are chosen within the range of base band of telephone channel (0 to 4000 Hz), then signalling is referred as voice frequency signalling. The band width of speech band occupies the range 300 – 3400 Hz. The selection of control signal frequency depend on different parameters. Voice frequency signalling is also referred as frequency division multiplex (FDM) system.

5.5.1 Inband Signalling

If the selection of control signal frequencies are within the speech band (300 to 3400 Hz) then the signalling is known as inband signalling. As the inband signalling freq. band is same as voice, the equipments used must be able to distinguish speech and signal. Some following parameters are available to do this task

(*i*) Selection of frequency for signal.

(*ii*) Signal recognition time.

(*iii*) Voice characteristics.

There are different pattern of voice frequency in amplitude and frequency changes. According to English language, the energy level of speech is predominent at lower frequencies and falls gradually over the rest of the region. To avoid noise, the selection of freq. is done in the range 2280 to 2600 Hz. Normal speech rarely produce a pure 2600 Hz, the adjacent frequencies distinguish it from the control signal.

The duration of the signals keep longer than the period for which the speech frequency is likely to period in speech, the change of signal limitation is reduced.

The probability of the signal receiver responding to a speech freq. can be reduced to a low level by a guard circuit as shown in Fig. (5.7). Here if the output from the signal circuit exceeds that from the guard circuit, the receiver operates and otherwise gives output signal.

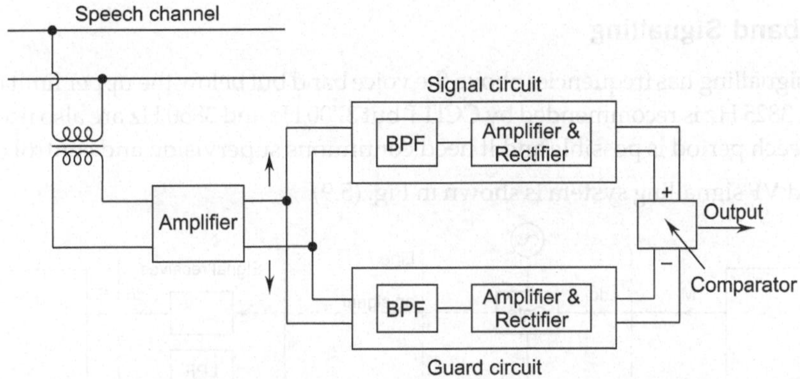

Fig. 5.7 *Voice frequency receiver*

The overall in band VF signalling system is shown in Fig. (5.8). Here, the line is split when the signal tone is transmitted in order to confine it to the link concerned and the line split at the end of the receiver is operated by the signal receiver. The buffer amplifiers are used to prevent transects produced by electromechanical switching equipment.

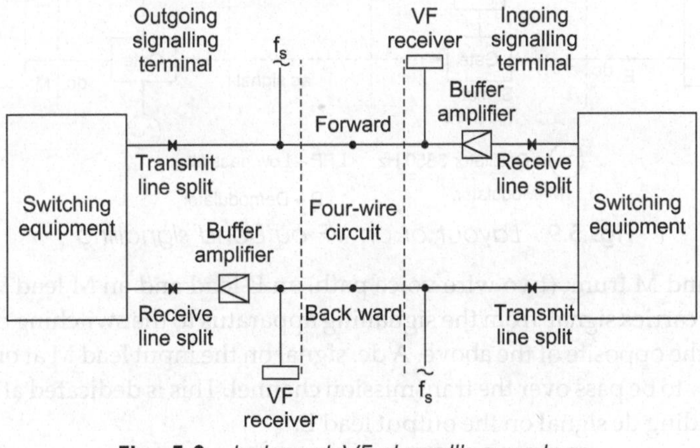

Fig. 5.8 *In-band VF signalling system*

Advantages

1. In band signalling is independent on the transmission system or medium.
2. Its operation is simple.
3. The control signals can be sent to every part where a speech signal can reach.
4. Inband signalling operation is flexible.

Disadvantages

1. There is a possibility of speech signal to limit the control signal guard circuit can reduce the problems
2. When multiple transmission links are connected end to end then the inband signal may "spill over" from one link to another, causing error in the system. Using line split to disconnect link whenever a signal is detected, can reduce the problem. Generally line split operates within 35 ms.

Application: Due to several advantages, the inband signalling system is used for long distance telephone networks.

5.5.2 Outband Signalling

The outband signalling has frequencies above the voice band but below the upper limit of 4 KHz. For his signalling, 3825 Hz is recommended by CCITT but 3700 Hz and 3850 Hz are also used. signalling during the speech period is possible and it need continuous supervision and control of a call.

An outband VF signalling system is shown in Fig. (5.9).

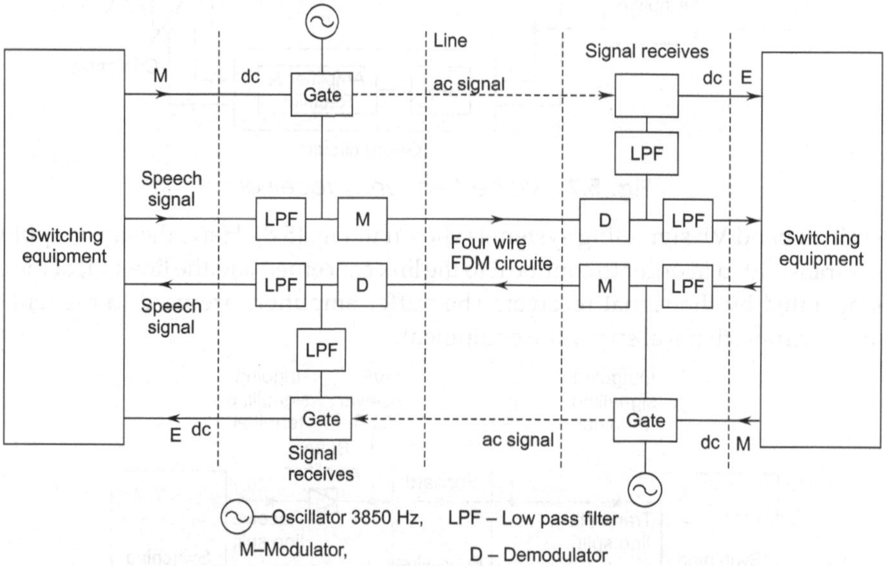

Fig. 5.9 *Layout of an VF outband signalling*

Here, 4-wire E and M trunk (two wire voice path, an E-load and an M lead with earth ground return). The E-lead carries signal from the signalling apparatus to the switching equipment and the M lead carries just the opposite of the above. A dc. signal on the input lead M at one terminal causes the signal frequency to be pass over the transmission channel. This is dedicated at the other terminal to give a corresponding dc signal on the output lead E.

If a route containing a section of audio frequency (AF), transmission or carries system not equipped with outband signalling, then it is necessary to use inband signalling.

Advantages

1. To avoid signal limitation, line splits are not required.
2. Signals and speech can be transmitted simultaneously without disturbing the conversation.
3. Outband signalling system is simple and cheap.

Disadvantages

1. Dependent on the transmission system.
2. In this signalling, the bandwidth available for signalling is very narrow.
3. For handling the bands, filter circuits are needed.

5.6 PCM SIGNALLING

In PCM signalling system, the signalling and speech are simply sampled, coded and transmitted within the frame of PCM channels in a convenient way. When the signalling and speech information carried in same time slot then it is referred as inslot signalling and for separate time slot it is referred as outslot signalling.

In the inslot signalling, the time slot is fixed at eight bits. As one bit is used for signalling, the speech bit rate is reduced to 56 Kbps from 64kbps and also the bandwidth is reduced. Using TDM, the telephone channels are combined to form an assembly of 24 or 30 channels, primary multiplex group. Two frame structures—DS1, 24 channel system and European 30 channel system are generally used.

(*i*) **DS1 24 channel system:** This system is also known as T1 (Telecommunication standards entity number 1) system. DSO is a 64 kbps signal that makes up the basis for the DS1. The structure of T1 link and DS1 frame format is shown in Fig. (5.10).

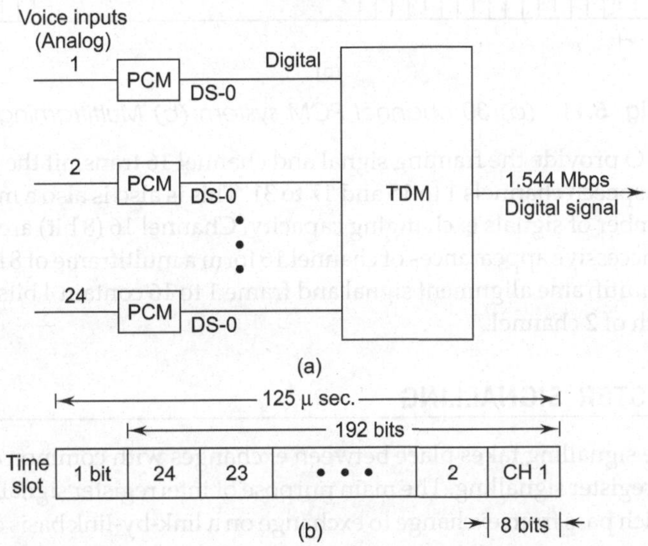

Fig. 5.10 *The structure of (a) T1 link and (b) DS-1 frame format*

Time division multiplexing is used for incoming analog signals and each TDM channel are assigned 8 bits per time slot and each frame contain 24×8 bits = 192 bis plus one additional bit to identify the frame boundaries. Thus, the total bits that each frame contains is 193 bits. The frame interval is 125 µsec and the basic T1 line rate is 1.544 Mbps, consider as the fundamental standard for digital transmission.

The old version of 24 channel system consist of 8 bit (7 speech bits and 1 signalling bit) but the latest version uses 12 frames together, referred as multi frame. In multiframe, every 6th frame within the 8 bit of each channel is used for signalling and 193nd bit of the PCM frame is used on alternate frames in multiframe arrangement to reduce the quantization distortion.

(*ii*) **30 channel PCM system:** in this system, there are 30 speech channels transmitted within a frame of 32 time slots. The total bit rate is $32 \times 8 \times 8$ kbits/sec = 2048 kbps. The Fig. (5.11) shows the 30 channel PCM system and multiframing structure.

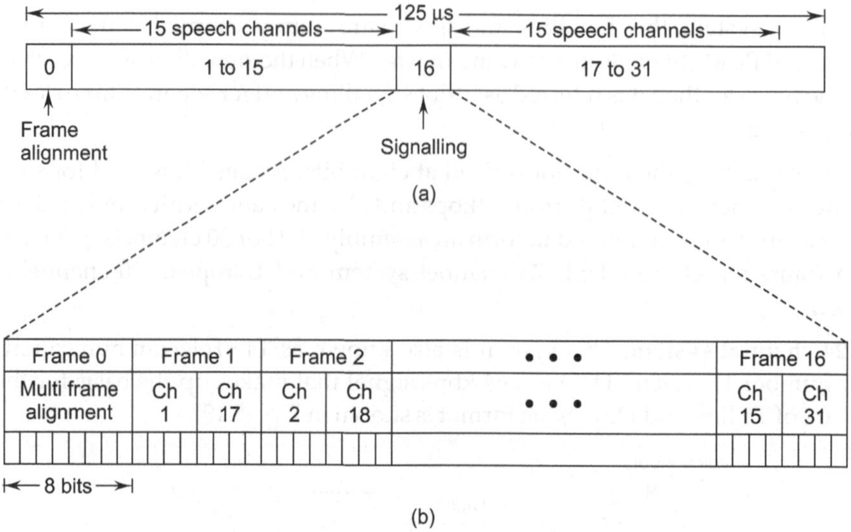

Fig. 5.11 *(a) 30 channel PCM system (b) Multiframing*

Here, channel O provide the framing signal and channel 16 transmit the signalling information related to speech channels 1 to 15 and 17 to 31. This is also is also a multiframing system with large number of signals exchanging capacity. Channel 16 (8 bit) are shared between 30 channels. 16 successive appearances of channel 16 form a multiframe of 8 bit time slots. Frame O contains a multiframe alignment signal and frame 1 to 16 contain 4 bits of signalling information for each of 2 channel.

5.7 INTER-REGISTER SIGNALLING

When interexchange signalling takes place between exchanges with common control subsystems then it is called interregister signalling. The main purpose of interregister signalling is the exchange of address digits which pass from exchange to exchange on a link-by-link basis or end-to-end basis.

Here, all the registers in the originating exchange receive address information from the called subscriber and send out routing digits in a successive manner until the terminating exchange is reached. Therefore, post dialling delay is produced. To minimize this delay, a rapid method of sending routing information than loop/disconnect pulsing is necessary and so inband multifrequency system have been developed.

Line signalling is also required in addition to inter-register signalling as inter register signalling system can not be used for seize, answer and clear signals. The Fig. (5.12) shows the link-to-link signalling where adjacent registers in a multilink connection, exchange the information.

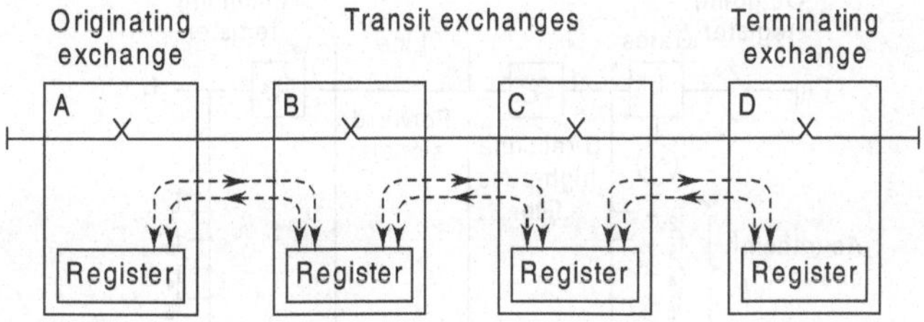

Fig. 5.12 *Link-by link signalling between Registers.*

Advantages

1. If the network being modernized all the registers need not to be modified as different signalling system may be used on different links.
2. Signals used only suffer from transmission impairments like noise of single link, distortion, attenuation.

End-to-end signalling system is shown in Fig. (5.13) below where an originating register controls the connection setup until it reaches its final destination. After this task, the connection is released and the originating register signals to the next register. Here each transit register have to select the outgoing route to the next telephone exchange by the received information.

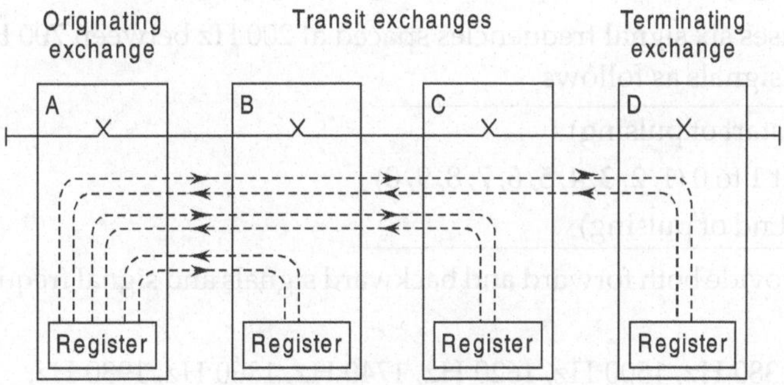

Fig. 5.13 *End-to End signalling between various registers or exchanges*

Disadvantage

1. All the registers must be compatible with the originating register.
2. Signals can suffer from transmission impairments of several links in tandem.

Advantage

1. Register holding time and post dialling delay are shorter than link-by link as each transmit register has to receive and transmit only part of the address information.

   In case of multifrequency inter register signalling system, each digit is sent by a combination of two frequencies out of six $\left(\frac{2}{6}\text{ MF}\right)$ with error detection capability and 15 possible digit values, five extra signals in addition to the digits '1' and '0'. The Fig. (5.14) shows the MF inter register system. The intervals between the two frequencies is 120 Hz or 200 Hz and VF receiver do not need speech immunity as d the system proceeds a conversation and short signal pulses are used.

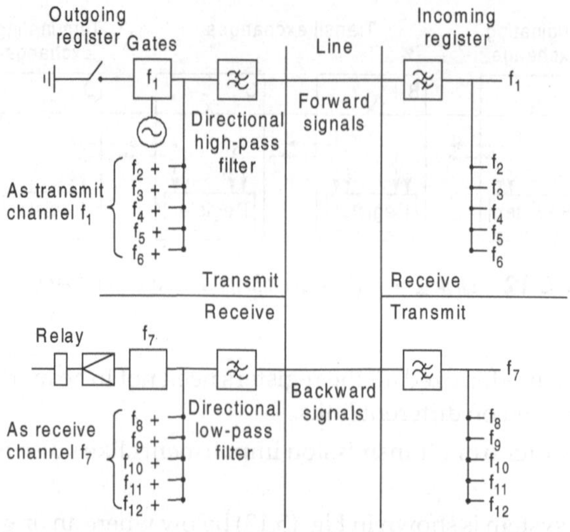

Fig. 5.14 *Multifrequency (MF) Inter-Register Signalling System*

The directional filters use two wire circuits for signalling

(*i*) Bell R1 signalling system and (*ii*) CCITTR2 signalling system.

The first system uses six signal frequencies spaced at 200 Hz between 700 Hz and 1700 Hz and transmit 12 forward signals as follows

KP (start of pulsing)
Digit 1 to 0 (1, 2, 3, 4, 5, 6, 7, 8, 9, 0)
ST (End of pulsing)

The second one provide both forward and backward signals and signal frequencies spaced at 120 Hz as follows,

Forward → 1380 Hz, 1500 Hz, 1620 Hz, 1740 Hz, 1860 Hz, 1980 Hz.

Backward → 540 Hz, 660 Hz, 780 Hz, 900 Hz, 1020 Hz, 1140 Hz.

Here more than 15 signals can be sent in each direction because shift signal (one signal frequency combination) is used to enable each of the others to have two meanings. So, that a wide repertoire of signals is available. Additional signals are used for international calls. Among the different national versions of the R_2 signalling system, British MF2 system is very poplar. Another version with pulse signalling is used for satellite circuits because of their long delay.

Note. Bell R_1 signalling system and CCITT R_2 signalling system known as CCITT Regional signalling system No. 1 and 2 respectively and the first one employ link by-link signalling and the second one based on end-to-end signalling system.

5.8 COMMON CHANNEL SIGNALLING (CCS)

This signalling technique use a separate common channel for passing control signals for a group of trunks or information paths. This channel is totally different from the one which carries the voice or data. Signalling in CCS is digital in nature, modems are used for carrying digital data over analog lines. The basic scheme is illustrated in Fig. (5.15).

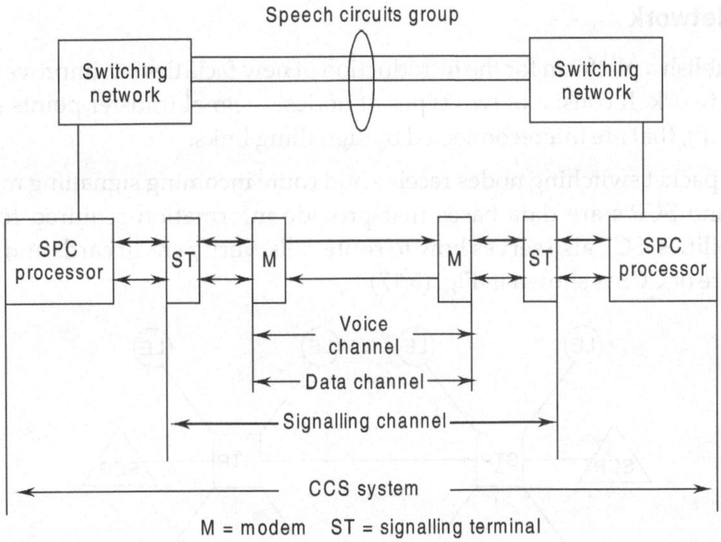

Fig. 5.15 *Basic scheme for CCS*

5.8.1 Basic Principle

The basic working principle of CCS depend upon the link-by-link basis along the route and store and forward network technique. In CCS, the data link sends messages that identify specific trunks and events. Two signalling channels, one for each direction are used in a dedicated manner to carry signal information so they are capable of carrying information for a group of circuit. CCS can carry signals for 1500 to 2000 speech circuits at the bit rate of 2.6 kbps. The CCS technique is also known as transparent mode for signalling.

The signalling information in CCS is transferred as messages of varying length usually defined as one or more fixed length signalling units (SUs). A message of one signal unit length is known as single unit message (SUM) and the one with multiple signal units is referred as multiunit message (MUM). A typical CCS signalling message format is shown in Fig. (5.16).

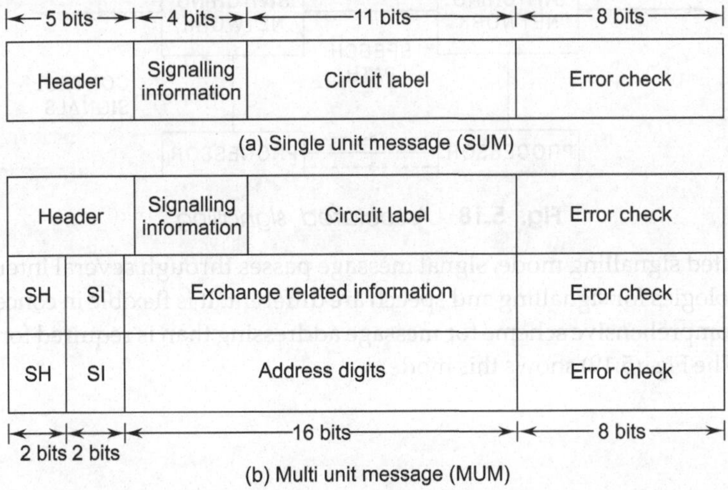

Fig. 5.16 *A typical CCS signalling message format*

5.8.2 CCS Network

CCS network establish a platform for the introduction of new facilities *i.e.,* improve the performance of the existing network. It consist of two types of nodes— signal transfer points (STP) and signal control points (SCP), that are intereconnected by signalling links.

The STP's, the packet switching nodes receive and route incoming signalling messages towards the destination and SCP's are data bases that provide information required for advanced call processing capabilities. SCP also serves how to route calls, verify credit cards and special services. The basic structure of CCS is shown in Fig. (5.17).

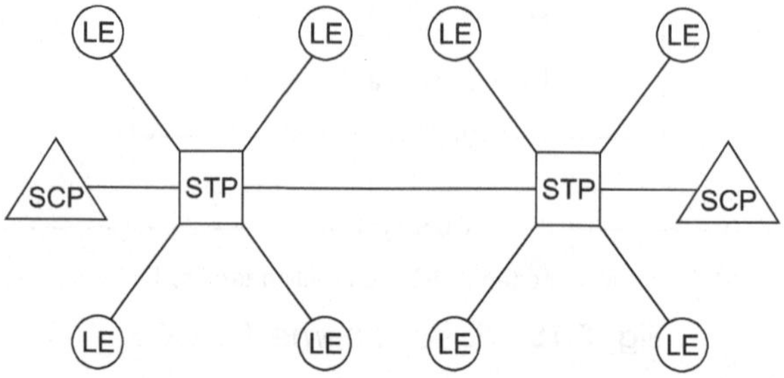

Fig. 5.17 *Basic structure of CCS*

There are three modes by which the CCS implementation is done

1. Channel associated mode.
2. Channel non-associated mode
3. Quasi associated mode

In channel associated signalling mode, a direct link is established between the two exchanges. The network topologies of signalling network and speech network are the same. It is simple, economic and easy to control. This enables an delay operation or long distance communication. The Fig. (5.18) shows the associated mode of signalling.

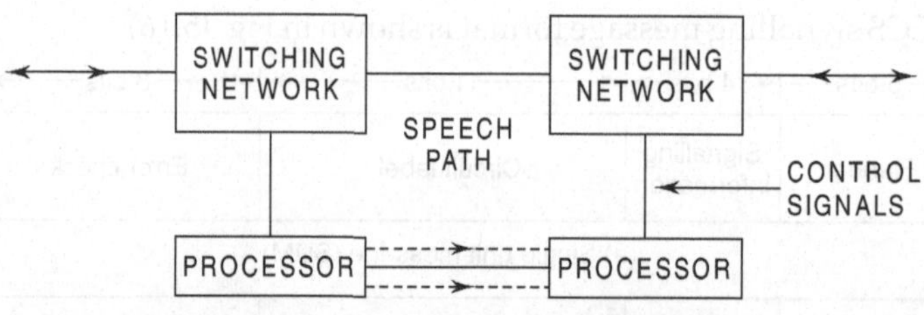

Fig. 5.18 *Associated signalling*

In non-associated signalling mode, signal message passes through several intermediate nodes. The network topologies for signalling and speech are different. It is flexible in concern of routing. It demands more comprehensive scheme for message addressing than is required for channel associated signalling. The Fig. (5.19) shows this mode.

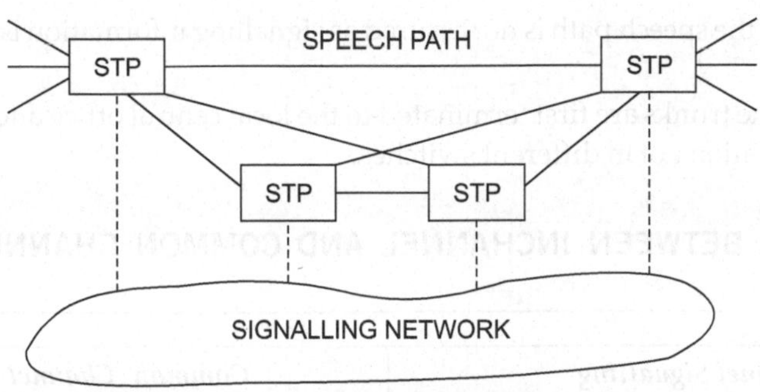

Fig. 5.19 *Non-associated signalling*

In quasi-associated signalling mode, a predetermined paths are established between exchanges. The message signals are passed through one intermediate node for short distance communication. For a given speech connection, the signalling paths are not associated but fixed. The Fig. (5.20) shows this mode.

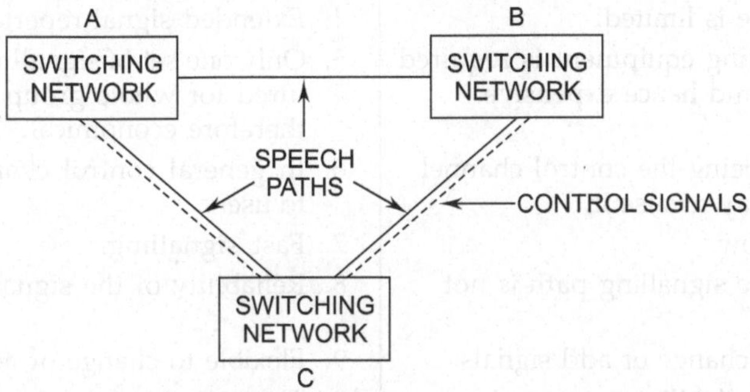

Fig. 5.20 *Quasi-associated signalling*

Advantages

1. CCS allows for signalling at any time within the entire duration.
2. Information can be exchanged between processors at high speed.
3. There is no chance of interference during transmission as separate channels are used for voice and control.
4. For each associated trunk group, only one set of signalling facility is required.
5. Efficient routing procedure and error rate is very low.
6. Cost is low as line signalling equipments are not necessary for every functions and inter-office trunk signalling cost can be removed.
7. CCS provides acceptable quality for network related signalling tones such as DTMF, MF and SF achieved.
8. Flexible to add more services.

Disadvantages

1. The CCS network works on S/F basis so causes additional overhead and disconnect the continuity.
2. If any node fails to transmit properly then there will be a question of reliability. So, high degree of reliability is required.
3. Proper inter facing facility is necessary as sometime CCS network is equipped with inchannel signalling.

4. The integrity of the speech path is not assured as signalling information is not sent over speech path.
5. With CCS, all the trunks are first terminated to the local central office and then forward to the different destination not in different switches.

5.9 DIFFERENCE BETWEEN INCHANNEL AND COMMON CHANNEL SIGNALLING

In Channel Signalling	*Common Channel Signalling*
1. Automatic propagation of signalling information enables the simultaneous process and release of associated facilities.	1. Signalling information must be relayed from one node to the next in store and forward (SF) fashion.
2. Trunks are held up during signalling	2. Trunks are not required for signalling.
3. Interface between voice and control signal may occur.	3. No interference occur because the channels are separated physically.
4. Signal repertoire is limited.	4. Extended signal repertoire is possible.
5. Separate signalling equipment is required for each trunk and hence expensive.	5. Only one set of signalling equipment is required for whole group of trunk circuit and therefore economical.
6. Voice channel being the control channel so customers may misuse it.	6. In general control channel is in accessible to users.
7. Signalling is slow.	7. Fast signalling.
8. Reliability of the signalling path is not critical.	8. Reliability of the signalling path is critical.
9. It is difficult to change or add signals.	9. Flexible to change or add signals.
10. Speech circuit reliability is assured.	10. There is no automatic test of the speech circuit.
11. During speech period it is difficult to handle signalling	11. There is freedom to handle signals during speech.
12. Transfer of information such as address digits is from common control network originating office— voice channel—receiving office—common control network.	12. Transfer of information is directly between control elements (processors of SPC systems).

5.10 CCITT SIGNALLING SYSTEM NO. 6 (SS6)

It is the first common channel signalling system to be standardized internationally. This system was designed for use in analog networks and used bit rates of 2.4 kbit/sec and 4.8 kbit/sec. In SS6, each SU is of size 28 bits of which the last eight bits are check bits. In SUM, 5, 4 and 1, 1 bits are allocated for header, signaling information and circuit label respecitvely. The signal units are grouped into blocks of 12, the last signal unit in each block being used for acknowledgment purposes. Each block contains a sequence number as part of the error detection and correction mechanism. Signal units within a block are either massage signal units or synchronization signal units. Within an exchange, the functions performed in transmission, CCITT signalling system no. 6. signal units are shown in Fig. (5.21).

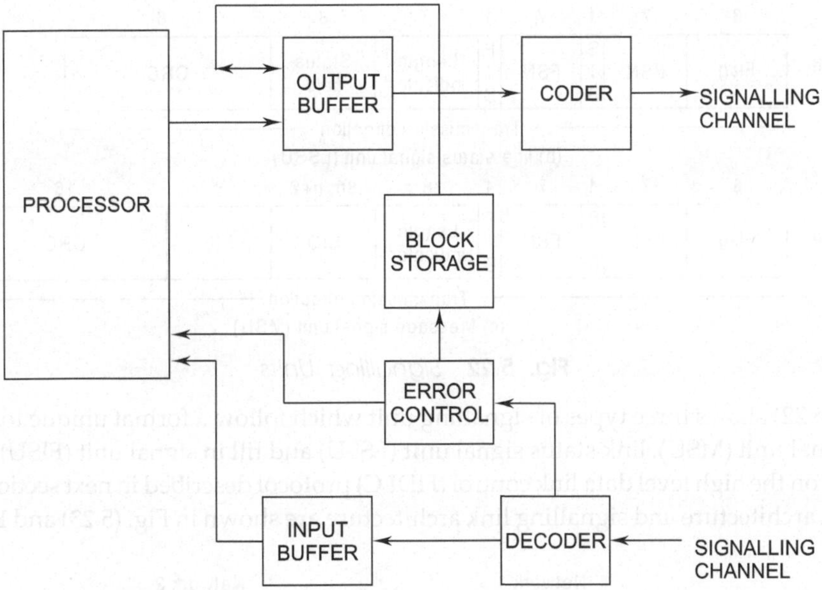

Fig. 5.21 *Exchange functions for CCITT no. 6 source*

In the functional block, the processor is responsible for call control within the exchange *i.e.*, conduct signalling communication. The processor pass the appropriate message to the output buffer which store each message until a proper slot is available for sending the message to the next exchange. The coder adds 8 check bits to the message to detect, the corruption.

At the receiving signal unit, the decoder checks each signal unit for errors with the 8 check bits. The signal unit is discarded in case of error found otherwise it is send to the input buffer which stores the signal unit until the processor analyse its contents. The corrupted signal unit may be retransmitted for detection. for this, a copy of each transmitted signal unit is stored in block storage until the receipt acknowledge of other exchange is received. Successful acknowledgement allows detection of signal unit stored in block storage otherwise retransmit the signal unit by the outgoing buffer.

5.11 CCITT SIGNALLING SYSTEM NO. 7 (SS7)

This is a global standard for telecommunication defined by ITU-T. SS7, in integration with SPC systems enables fast and efficient call setup and release accross the network in less than one second. Basically, SS7 use the concept of packet switching with common channel signalling technique. The signalling unit of SS7 is shown in Fig. (5.22).

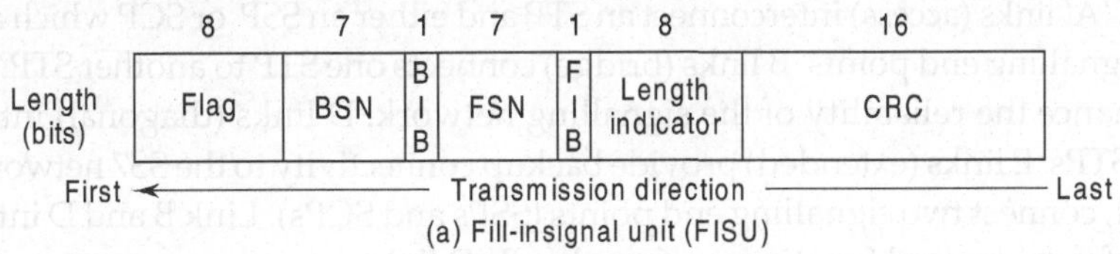

(a) Fill-insignal unit (FISU)

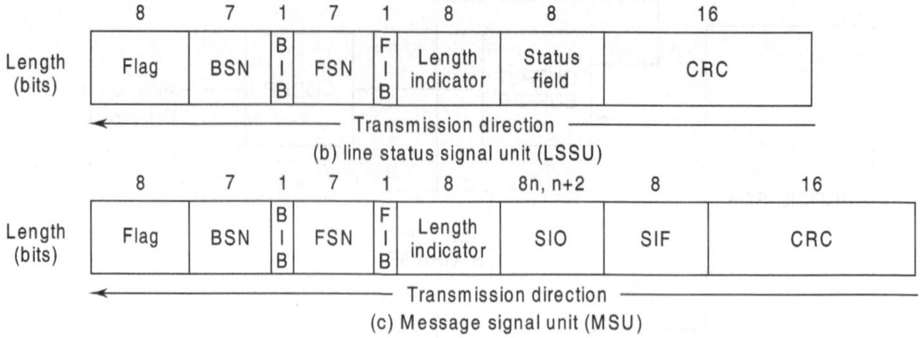

Fig. 5.22 *Signalling Units*

The Fig. (5.22) shows three types of signalling unit which follow a format unique to that type—Message signal unit (MSU), link status signal unit (LSSU) and fill in signal unit (FISU). The signal unit is based on the high level data link control (HDLC) protocol described in next section. A typical SS7 network architecture and signalling link architecture are shown in Fig. (5.23) and Fig. (5.24).

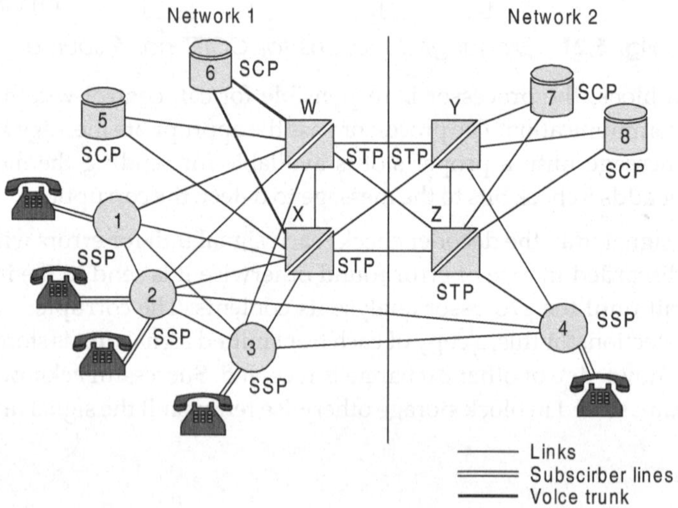

Fig. 5.23 *SS7 network architecture*

The SS7 network is built out of the essential components interconnected by signalling links. The messages are exchanged between network elements over 56 or 64 kbps bidirectional channels called signalling links.

In Fig. (5.24), 'A' links (access) interconnect an STP and either an SSP or SCP which are collectively referred to as signalling end points. B links (bridge) connects one STP to another STP. C links (cross) are used to enhance the reliability of the signalling network. D links (diagonal) interconnect two mated pairs of STPs. E links (extended) provide backup connectivity to the SS7 network. F stands for fully associated, connect two signalling end points (SSPs and SCPs). Link B and D interconnect two mated pair (perform identical functions) referred as B/D links.

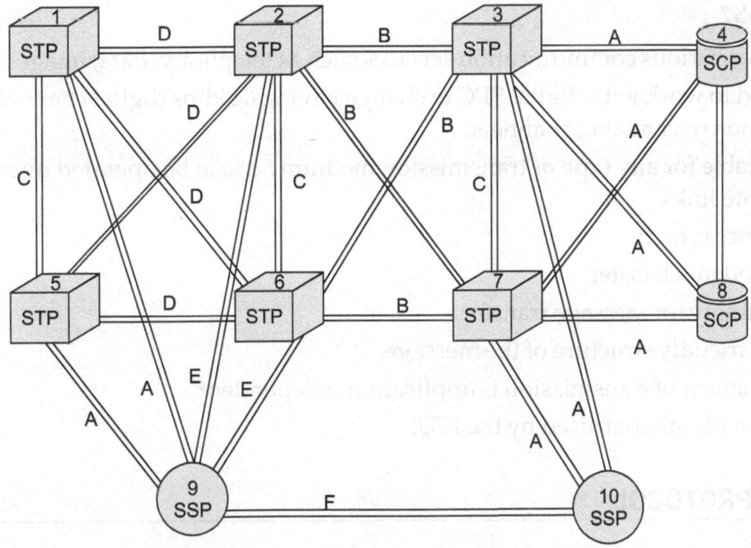

Fig. 5.24 *Signalling Link architecture*

The block diagram of SS7 system is shown in Fig. (5.25). Here the signal messages are transmitted to the CCS system from the central processor of the sending exchange. This system based on three microprocessor subsystems

(*i*) Signalling control subsystem

(*ii*) Signalling termination subsystem

(*iii*) Error control subsystem.

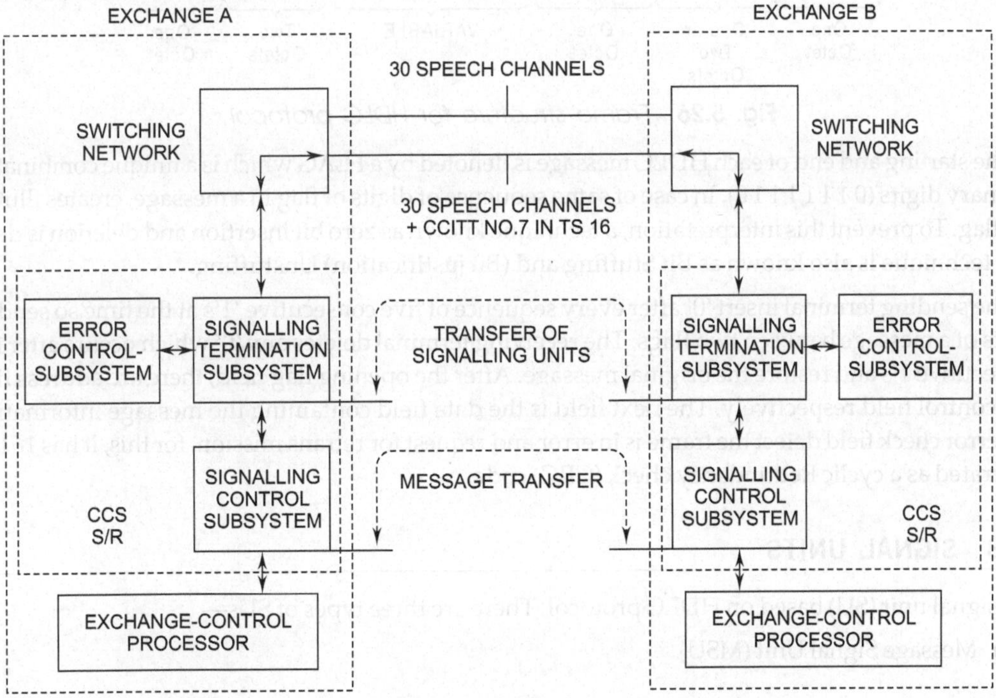

Fig. 5.25 *Block diagram of CCITT no 7 signalling system*

Features of SS7

1. Suitable for various communication services such as telephony, data, images and video.
2. SS7 is used to work with digital SPC exchange having 64 kbps digital channels, also suitable for operation over analog channels.
3. SS7 is suitable for any type of transmission medium *i.e.*, can be operated over both terrestial and satellite links.
4. Performance is high.
5. Flexible and much faster.
6. High reliability for message transfer.
7. Processor friendly structure of the message.
8. The mechanism of transmission is application independent.
9. Internationally standardised by the ITU.

5.12 HDLC PROTOCOL

CCITT signalling system no. 7 uses the International Standard protocol, HDLC (High Level Data Link Control). The frame structure for HDLC protocol is shown in Fig. (5.26). The frame contain the number of packets of a message.

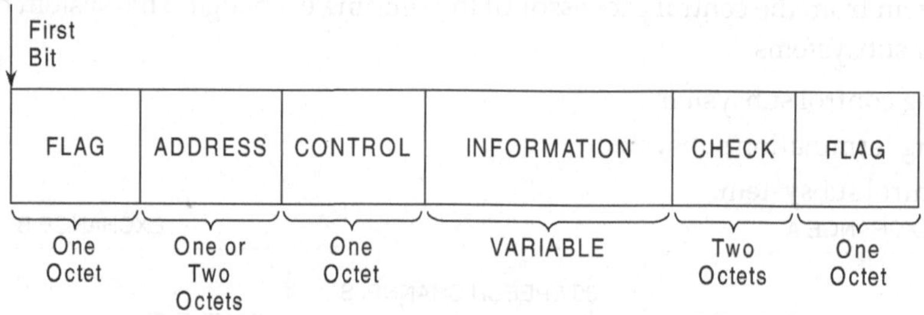

Fig. 5.26 *Frame structure for HDLC protocol*

The starting and end of each HDLC message is denoted by a FLAG which is a unique combination of binary digits (0 1 1 1 1 1 1 0). In case of same sequence of digits of flag in a message, creates illution as a flag. To prevent this interpretation, a technique known as zero bit insertion and deletion is done. This technique is also known as Bit Stuffing and (Bit justification) Unstuffing.

The sending terminal insert '0' after every sequence of five consecutive '1's at the time so sending digits of a message between two flags. The receiving terminal deletes any '0' which occurs after five consecutive 1's and restore the original message. After the opening flag field, there are address field and control field respectively. The next field is the data field containing the message information. The error check field detect the frame is in error and request for retrans mission. for this, it has 16 bits, generated as a cyclic Redundancy check (CRC) code.

5.13 SIGNAL UNITS

The signal unit (SU) based on HDLC protocol. There are three types of SUs—

(*i*) Message Signal Unit (MSU)

(*ii*) Link-status Signal Unit (LSSU)

(*iii*) Fill-in-Signal Unit (FISU).

The Fig. (5.22) shows the format of three types of SUs in section (5.11). Now we will discuss the three types in detail.

- Message Signal Unit (MSU) transfers information supplied by a user part (Level 4) via Signalling Network Level (Level 3).
- Link Status Signal Unit (LSSU) is used for link initialization and flow control.
- Fill in Signal Unit maintain alignment when there is no signal traffic.

The fields used in all the signalling units are described below:

(*a*) **Flag:** The indication and working of flag field is same as in HDLC protocol frame format explained in section (5.11).

(*b*) **Backward Sequence Number (BSN):** This field is used to acknowledge the receipt of signal units by the remote signalling point. It contains the acknowledged signal units sequence number.

(*c*) **Backward Indicator Bit (BIB):** It's a negative acknowledge, indicated by inverting the BIB bit.

(*d*) **Forward Sequence Number (FSN) and Forward Indicator Bit (FIB):** FSN contain the sequence number or identification of the signal unit using module 128 count. It is incremented by 1, when the SU is ready for transmission. FIB is used in error recovery like the BIB.

(*e*) **Cyclic Redundancy Check (CRC):** The value of CRC is used to detect and correct data transmission errors. It contain 16 bits.

(*f*) **Length Indicator (LI):** This field indicates the number of octets (8 bit bytes) between itself and CRC check sum *i.e.*, it gives the length of the signal unit. The 6 bit LI can store values between zero and 63. A value of LI greater than two referred the SU as message signal unit. The maximum length of a signal unit is 279 octets— 273 octets (data) + 1 octet (flag) + 1 octet (BSN + BIB) — 1 octet (FSN + FIB) + 1 octet (LI + 2 BITS SPARE) + 2 OCTET (CRC).

(*g*) **Service Information Octet (SI0):** This field remain only in MSU. It contains 4 bit sub-service field followed by 4 bit service indicator. The sub-service field contains the network indicator (*i.e.*, national or international) and the message priority. The service indicator specifies the MTP (Message Transfer Part) user, thereby allowing the decoding of the information contained in the SIF.

Service Indicator	MTP User
0	Signalling Network Management message (SNM)
1	Maintenance Regular Message (mtn)
2	Maintenance Special message (MTNS)
3	SCCP (Signalling Connection Control Part)
4	TUP (Telephone user part)
5	ISUP (ISDN user part)
6	Data user part (call and circuit related messages)
7	Data user part (facility registration/cancellation messages)

(*h*) **Signalling Information Field (SIF):** This field is also in MSU only. It contains the information to be transmitted. SIF may contain upto 272 octets. Detailed description is in the next section. LSSU and FISU are more simple than MSU.

5.14 SIGNALLING INFORMATION FIELD (SIF)

A message from the user part to be transmitted occupies the SIF. Field. An example of telephone user part (TUP) is the Initial Address Message (IAM) which is the first message to be sent.

The format of IAM in CCITT signaling system no. 7 is shown in Fig. (5.27).

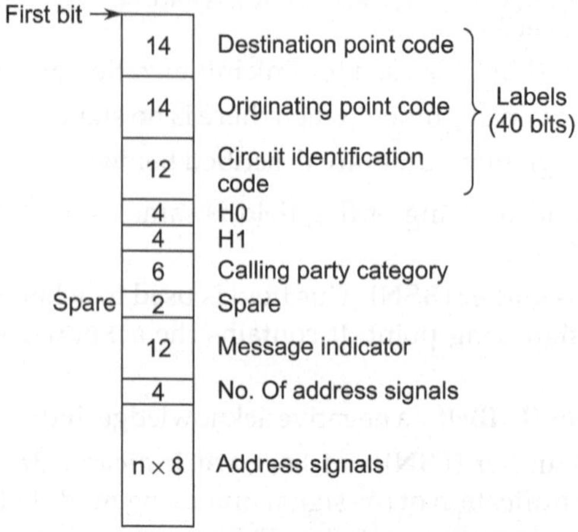

Fig. 5.27 *Format of IAM in CCITT signalling system no 7 (SS7) system*

The sequence of transmission of various field of IAM is as follows—

1. A leble of 40 bits contains:
   - The Destination Point Code (DPC) — 14 bits
   - The Originating Point Code (OPC) — 14 bits
   - The Circuit Identity Code (CIC) — 12 bits
2. The HO/H1 Octet:
   - The HO field (4 bits) identifies a general category of message (e.g., forward address message)
   - The H1 field completes the definitions of the message (e.g., variable length IAM).
3. The Calling Party Category (e.g., Ordinary customer or operation).
4. Message Indicator: Any special requirements are indicated by this field (e.g., whether a satellite link or echosuppressor will be used).
5. The number of address signals (4 bits): This field indicates the number of address digits in IAM.
6. The address signal:

The SIF then followed by the check bits and closing flag to complete the MSU. In short, we can say the SIF transmit the information following the routing label.

SUMMARY

- A signalling system link the various switching systems, transmission systems and subscriber equipments in a telecommunication network to perform efficiently as a whole.
- Three types of signalling — subscriber loop signalling, register signalling and interregister signalling are involved in telecommunication network.

- Interregister signalling exchange the address digits based on link by link basis or end to end basis.
- The technique of signalling is classified into two classes (*i*) inchannel signalling and (*ii*) common channel signalling.
- Inchannel signalling use the same channel that carries user voice or data to pass control signals related to that call or connection.
- Inchannel signalling is sub divided into two groups (*i*) inband signalling and (*ii*) outband signalling.
- Inband signaling use the same frequency band as the voice band *i.e.*, 300 Hz to 3400 Hz.
- Outband signalling uses frequencies above the voice band but below the upper limit of 4000 Hz or 4 kHz.
- Common channel signalling uses a dedicated line for signalling information.
- CCITT signalling system no 6 and 7 are the examples of common channel signalling.
- CCITT no. 6 system was designed for use in analog networks and used bit rates of 2.4 kbit/sec and 4.8 kbit/sec.
- CCITT no. 7 system is a three processor based system.
- HDLC protocol is used by CCITT no. 7 system.
- The signalling information is transferred as message (variable length) usually defined as one or more fixed length signalling units (SUs).
- There are three types of SUs
  - (*i*) Message signal unit (MSU)
  - (*ii*) Link-status signal unit (LSSU)
  - (*iii*) Fill in signal unit (FISU).

REVIEW QUESTIONS

1. Define signalling. Why it is required?
2. What do you mean by customer line signalling? Discuss.
3. Make the classification chart of signalling technique.
4. Discuss the principle of operation of PCM signalling.
5. Define inchannel signalling.
6. Briefly discuss inband and outband signalling. List the advantages and disadvantages of both the signalling.
7. What do you mean by common channel signalling? Discuss its merits and demerits.
8. Compare the two types of signalling technique.
9. What do you mean by inter-register signalling? Explain.
10. Explain CCITT signalling system no 6. What is the difference between CCITT 6 and 7 system?
11. Discuss HDLC protocol.
12. Describe signal unit.

MULTIPLE CHOICE QUESTIONS

1. Inband voice frequency uses the frequency band

   (*i*) 30 – 300 Hz (*ii*) 300 – 3000 Hz

   (*iii*) 300 – 3400 Hz (*iv*) 3 – 340 Hz.

2. CCITT signalling system no 6 uses the bit rate of

   (*i*) 3.6 kbit/sec (*ii*) 4.8 kbit/sec

   (*iii*) 2.4 kbit/sec (*iv*) both (*ii*) and (*iii*)

3. SIF is the field of
   (*i*) MSU   (*ii*) FISU
   (*iii*) LSSU   (*iv*) both (*ii*) and (*iii*)

4. Inchannel signalling is also known as
   (*i*) DC signalling   (*ii*) Subscriber loop signalling
   (*iii*) Per trunk signalling   (*iv*) None of the above.

5. Only one set of signalling equipments is required for a whole group of trunk circuits—
   (*i*) Common channel signaling   (*ii*) Inchannel signalling
   (*iii*) Inband signalling   (*iv*) Outband signalling

6. In inchannel signalling both ac and dc signalling is done
   (*i*) True   (*ii*) False

7. In which technique of signalling speech circuit reliability is assured
   (*i*) Inchannel signalling   (*ii*) Common channel signalling
   (*iii*) None of the above   (*iv*) both (*i*) and (*ii*)

8. Common channel signalling based on—
   (*i*) Link-by link signalling   (*ii*) End to end signalling
   (*iii*) None

9. CCITT signalling system no 7 can be modelled same as—
   (*i*) TCP/IP model   (*ii*) ISO model
   (*iii*) OSI model   (*iv*) None of the above.

10. Exchanges are designed to accept a maximum loop resistance of—
   (*i*) 100 Ω   (*ii*) 1000 Ω
   (*iii*) 1300 Ω   (*iv*) 1500 Ω

ANSWERS

1. (*iii*)	2. (*iv*)	3. (*i*)	4. (*iii*)	5. (*i*)
6. (*i*)	7. (*i*)	8. (*i*)	9. (*ii*)	10. (*iii*)

6

Packet Switching

6.1 INTRODUCTION

In recent world packet switching is most popularly used technology as it is related to computer network and overcame all the limitations of circuit switching and message switching (described in chapter 1). Basically packet switching is highly suitable for the data communication. A long messages in the packet switching need control over it. Therefore for controlling of hardware implementation to reduce traffic and delay time a system known as ATM (Asynchronous Transfer Mode) is designed. The ATM concept is described in the section (6.10).

6.2 PACKET SWITCHING

In packet switching, the data or message is divided into packets of fixed or variable size for transmission through the network. Each packet transmission is independent of the other and the time interval between consecutive packets may vary. Therefore each packet need to carry complete address information *i.e.*, source id, destination id, packet id, along with the user data.

As the bits in a packet arrive at a switch or router, they are read into a buffer. After storing the entire packet, the router routes the packet over one of its outgoing link. The packets remains queued in the buffer until the outgoing link becomes ideal. The working principle of packet switching is shown in Fig. 6.1.

From the figure (6.1) it is seen that all the packets of a single message transmit from source to destination via different path following the store and forward technique and depending upon the availability of the free link/idle link.

In packet switching network, there are two factors on which delay depends

(*a*) Storage delay, T_S

(*b*) Forward delay, T_f

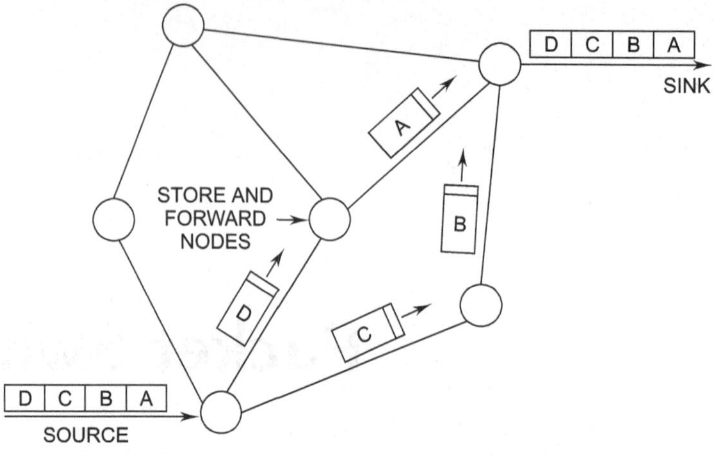

Fig. 6.1 *Packet switching network*

Therefore store and forward delay, $\boxed{T_{sf} = T_S + T_f}$ ...(6.1)

Now, T_f may be expressed as,

$$\therefore \quad \boxed{T_f = (n-1)(T_q + T_m) + T_t} \quad ...(6.2)$$

where, N = No. of nodes involved in the transfer.

T_q = Average queuing delay at each node.

T_m = Processing delay in each node

T_t = Message or packet transmission delay.

If the transfer rate (R) is uniform on all the links, then the data transmission delay is given by,

$$\boxed{T_t = (N+1)\frac{M}{R} + T_p} \quad ...(6.3)$$

where, M = Message length

T_p = Propagation delay from source to destination.

If the data transfer rate is non-uniform then T_t is given as —

$$\boxed{T_t = M\left[\frac{1}{R_1} + \frac{1}{R_2} + ... + \frac{1}{R_{N+1}}\right]} \quad ...(6.4)$$

Basically, transmission delay remain constant irrespective of the load.

6.2.1 Packets Format

In packet switching, all the messages are divided into packet form and transmit from source to destination in the form of packet. So, the format of the packet is very important (format can vary from one network to another). The fig. (6.2) shows the typical packet format.

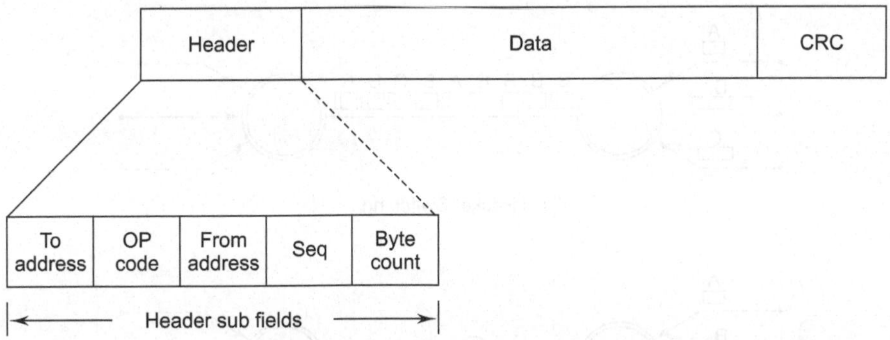

Fig. 6.2 *Typical packet format*

The packet format is divided into three part

(*i*) Header (*ii*) Data (*iii*) CRC (Cyclic Redundancy check)

(*i*) **Header:** Header includes the necessary address fields (source address and destination address) and relate control information fields. Generally three control information fields are

(*a*) *Op-code :* It indicates whether the packet is message (text) packet or control packet.

(*b*) *A sequence number (Seq) :* Detect faults and provide recovery facilities procedure.

(*c*) *Byte Count :* Indicate the size or length of the packet.

(*ii*) **Data :** It contains a portion of data/message. If the packet is for control purpose then there will be no data/message in this field.

(*iii*) **CRC :** This field specify some small numbers of errors with the help of a set of parity bits.

6.3 STATISTICAL MULTIPLEXING

In care of digital time division multiplexing there are two types of multiplexing—

(*i*) Synchronous Time Division Multiplexing (STDM) and

(*ii*) Asynchronous Time Division Multiplexing (ATDM).

The ATDM is known as Statistical Time Division Multiplexing (STATMUX), This statistical multiplexing is related to the packet switching techniques. The commercially available statistical multiplexers can be used to build up a packet switching network.

The format of the frame used by statistical multiplexer is identical to synchronous time division multiplexer. The main difference is that STATMUX periodically redefines the length of its frame to change the member of time slots or the number of channels whereas STDM permanently assign the time slot for each source. Whenever the source becomes active, time slot is assigned by STATMUX. Each time slot of STATMUX system is shorter and contains only source data.

The purpose and performance of statistical multiplexing systems are similar to the packet switching link. If the size of the packet is small then the operation of STATMUX and packet switching is same. The basic operation of packet switching and statistical multiplexing is shown in figure (6.3).

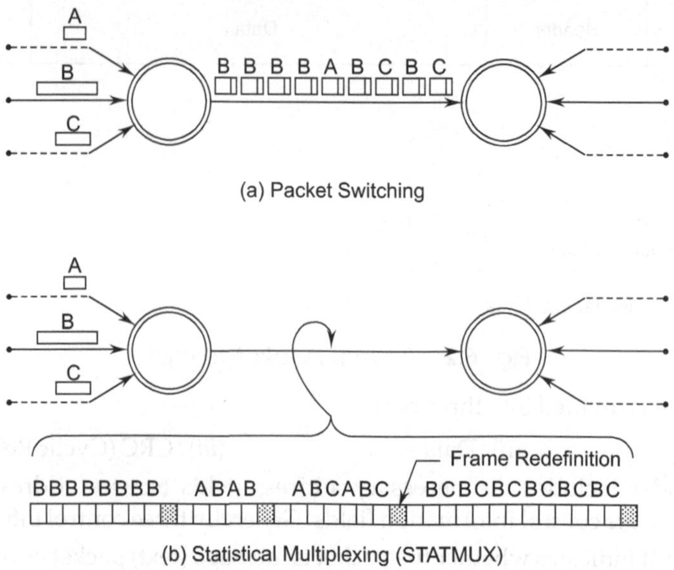

Fig. 6.3 *Illustration of comparison of packet switching and statistical multiplexing (STATMUX)*

As shown in the figure above the statistical multiplexing breaks the message upto even finer blocks (words) or data and adds periodic frame definitions messages so that the receiving terminals can identify the individual time slots and switch the incoming data properly. Error control functions are redundancy checked and request for transmission.

So, we conclude that with the help of this statistical multiplexing the packet switching network perform its operation smoothly.

6.4 ROUTING CONTROL

It is known that, in packet switching system, messages are broken into packets and sends to the network one after another. it is the routing control that decide how the network will handle the stream of packets to route them from source to destination through the network. There are two ways to determine the decision of the routing

(*i*) Datagram and (*ii*) Virtual Circuit.

6.4.1 Datagram

Datagram is the independent routing method of each packet within a stream. There is a routing table (may be static or periodically updated) stored in the router (switch), indicate the outgoing link for each destination. In the another case, router's estimate of the shortest path to the destination decide the routing. Each packet contain source and destination bits as each packet may be routed over different links. The concept of this method will be cleared with the help of the fig. (6.4).

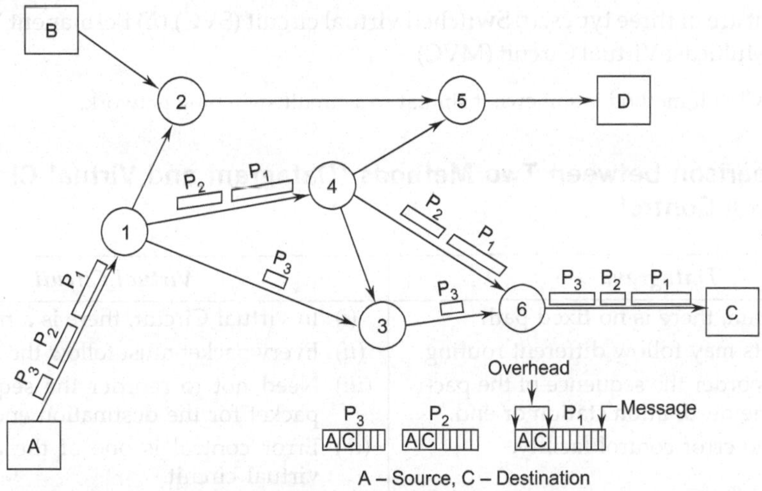

Fig. 6.4 *Basic concept of datagram*

In the fig. (6.4) it is seen that three packets P_1, P_2 and P_3 have to send from source terminal (A) to destination or end terminal (C) over the network. The source (A) sends these packets to the nearest node (provide switching facility) 1 which decides further routing. Node 1 select node 4 as a shortest path so that transmit P_1 and P_2. In case of not accessible node 4, another nearest path node 3 on which P_3 packet is send from node 1. Now, node 3 and 4 transmits the received packets to the destination through node 6. As the routing of each packet is different so the sequence of packets become disordered. The destination end take the responsibility to make the sequence of the packets and to recover the lost packet.

6.4.2 Virtual Circuit

In case of virtual circuit, a fixed path is established between source and destination over which the stream of packets are transmitted. All packets belonging to the same data stream follow this fixed route called a virtual circuit. All the packets contain a virtual circuit identifier (usually shorter than the overhead bits of the datagram).

The Fig. (6.5) shows the concept of virtual circuit. Here source B sends a control packet referred as call-request packet to node 2, requesting logical connection to D. Similarly node 2 decide the route and send the call-request to node 3, node 4 and to destination D. If the destination D is ready to establish the connection then, it sends a call-accept packet to node 4 and node 4 forward it to the source B through node 3 and 2. Each data packet with virtual circuit identifier and data transmit from B to D. There is no need to reorder the packets as they follow a fixed path.

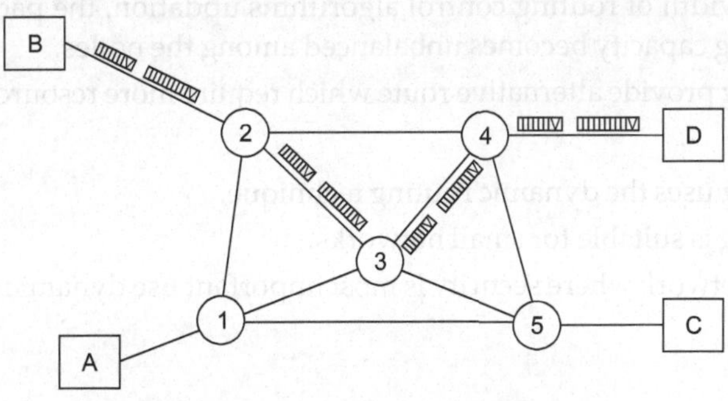

Fig. 6.5 *Concept of virtual Circuit*

Virtual circuit are of three types: (*i*) Switched virtual circuit (SVC) (*ii*) Permanent Virtual Circuit (PVC) and (*iii*) Multicast Virtual Circuit (MVC)

Note. Virtual Circuit method is somewhat similar to a circuit switching network.

6.4.3 Comparison between Two Methods (Datagram and Virtual Circuit) of Routing Control

Datagram	*Virtual Circuit*
(*i*) In Datagram, there is no fixed path.	(*i*) In virtual Circuit, there is a fixed path
(*ii*) All packets may follow different routing.	(*ii*) Every packet must follow the same routing.
(*iii*) Need to reorder the sequence of the packets for the destination station or end.	(*iii*) Need not to reorder the sequence of the packet for the destination end.
(*iv*) There is no error control facility.	(*iv*) Error control is one of the advantage of virtual circuit.

6.4.4 Dynamic Routing

Dynamic routing based on the distribution of network nodes. It provide the appropriate outgoing link after examining the distribution address of each received packet. The selection of outgoing link depend on the locally stored information that which path provides minimum delay to the distribution. The individual packets of a single message may follow different paths though the network as the decisions of routing are rapidly adjusted. To maintain this adjustment, the routing criteria is adjusted time to time to include both the operational status and size of the queues in the adjacent nodes.

Advantages

1. It has the ability to respond quickly to changes in the network topology or traffic condition.
2. Dynamic routing ensure the survivability of the network.

Disadvantages

1. The packets of a message may arrive at the destination out of sequence as the dynamic routing allow the packets to follow different path. So that there is a need to rearrange the sequence number of different packets which is very complicated.
2. Dynamic routing has the possibility of oscillations occurring in the routing decisions.
3. Dynamic routing present difficulties in proper distribution control.
4. For small bandwidth of routing control algorithms updation, the packets might lost as the traffic controlling capacity becomes unbalanced among the nodes.
5. Dynamic routing provide alternative route which require more resources.

Application

(*i*) Packet switching uses the dynamic routing technique.
(*ii*) Dynamic routing is suitable for small networks.
(*iii*) In the military network where security is most important use dynamic routing.

6.4.5 Fixed Path Routing

Fixed path routing works same as virtual circuit networks except successive connections, or sessions between any two end points always use the same path. Basically, fixed path routing assigns permanent virtual circuits to each pair of end points. Another exception is that in fixed path routing, there is no call establishment phase but it sends "clear to transmit" message before transmission and after the required resources are permanently allocated to become active. Therefore, fixed path routing provide two modes of operation.

(*i*) Permanently Active (Hot line) virtual circuits.

(*ii*) Virtual circuits Activated and Deactivated as required.

Advantages

1. In the fixed path routing, there is no successive connections i.e., circuits complicacy is minimum.
2. Call establish function is absent in the fixed path routing.

Disadvantages

1. The possibility of node or link failure is high in the fixed path routing.
   This problem can be avoided by using network control centre.
2. Fixed path routing is implemented with some older message switching network in the strict sense.
3. This type of routing is suitable for message switching network as delivery is less time critical.

Note. The selection of appropriate path for the transmission of messages from source to destination is known as routing. Depending on datagram and virtual circuit concept, virtual, dynamic and fixed path routing are also used for different type of networks.

6.4.6 Virtual Circuit Routing

This type of routing based on the virtual circuit concept which is discussed previously in section (6.4.2). To control the routing, it establish the virtual circuits in two ways

(*i*) Distributed control and (*ii*) Centralized control.

In case of distributed control, each node has the right to take the local decision to select the outgoing link. In the centralized control wide view of network status and traffic patterns are studied to select the outgoing link.

Advantages

1. Virtual Circuit Routing control the transmission in such a way that all the packets are transmitted in a orderly manner. There is no need to rearrange the sequence of the packets at the destination.
2. This routing mainly works in packet switching network.
3. The address requirement of each packet is reduced.
4. It has its inherent ability to aid in flow control.

Disadvantages

1. There is a possibility of greater transmission delay.
2. In case of variation of traffic pattern, packets to a particular virtual circuit may have to face long queuing delays while alternative links are lightly loaded.
3. In actual practice, virtual circuit has no dedicated transmission facilities.

6.5 FLOW CONTROL

In communication system, control is one of the important feature to handle the traffic. This requirement is applicable for any type of networks like packet switched, circuit switched and message switched network. There are basically two types of control

(*i*) Routing control and (*ii*) Flow Control.

In this section we will discuss about flow control in packet switched network. In case of congestion, new service requests must be blocked at the periphery of the network before they tie up common resources (store and forward buffers and transmission links in packet switching) and congestion will increase. In packet switching network, the flow control is concerned with buffer management. The buffer transmit the packet when an acknowledgement is returned from the adjacent nodes If the adjacent nodes are filled with packets then they are unable to receive additional packets and can not acknowledge. Therefore flow control requirements need that all the nodes are aware of traffic and in case of overload, refuse new requests until the congestion is over. If virtual circuit is used in packet-switching, then there is a possibility that the entire request will be serviced in a perfect manner with time. But when the packet switching network already faces the overloaded condition, attemps to establish virtual circuits might be suspended.

Flow control within the network is automatically established within each virtual circuit. Some virtual circuits contain large number of buffers compare to others, allow high data rate by accepting frequent clear to transmit signals whereas other nodes within other virtual circuit adjust themselves automatically to the maximum rate that the source and destination can support. If the allocated buffers of the participating nodes are full then the source stop sending packets as the condition propagates back to the source. This operation is known as back pressure.

When the network links are overloaded then the bandwidth of the flow control is minimal. Moreover, packet transmission stops if the flow control signals stop. Therefore, for smooth transmission flow control is very important.

Note. Flow control refers to a set of procedures used to restrict the amount of data the sender can send before waiting for acknowledgement.

6.6 X.25 PROTOCOL

X.25 is an international standards organization related to united nations (ITU-T standard) that defines the interface between a Data terminal equipment (DTE) and a packet switching network through a data circuit terminating equipment (DCE) or MODEM.

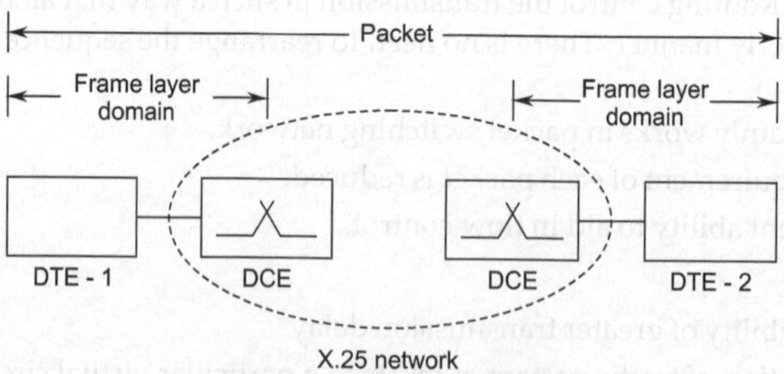

Fig. 6.6 *X.25 Interface*

The X.25 protocol is a packet-switched virtual circuit network created at the network layer. It defines, how a packet-mode terminal can be connected to a packet network for exchanging data.

It use to provide necessary procedures for establishing, maintaining and terminating connections and also gives some facilities like reverse charge, call direct and delay control. It is also known as subscriber network interface (SNI) protocol.

The network level formatted packet is shown in fig. (6.7).

Flag	Address	Control	GFI	LCGN	LCN	PTI	Data	Frame check	Flag

Fig. 6.7 *Formatted packet in network level*

where, GFI— General Formal Identifier
LCGN— Logical Channel Group Number
LCN— Logical Channel Number
PTI— Packet type Identifier
Flag— (*i*) Parity flag
(*ii*) Zero flag
(*iii*) Sign flag
(*iv*) Carry flag

The control fields of the network packet identify the type of packet in use. It is noted that, X.25 standard is applied where there is no controlling primary station available in network.

Note. The X.25 protocol specifies three layers — physical layer, frame layer and packet layer.

6.7 FRAME RELAY

Today's LAN and computing equipment have the potential to run at much higher speeds and transfer very large quantities of data. Frame relay which is a new Wide Area Networking (WAN), is going in popularity. It uses a packet switching technology or virtual circuit technology, similar to X.25 but is more efficient. As a result, it can make the networking quicker, simpler and less costly. A schematic diagram of a typical public frame relay network is shown in Fig. (6.8).

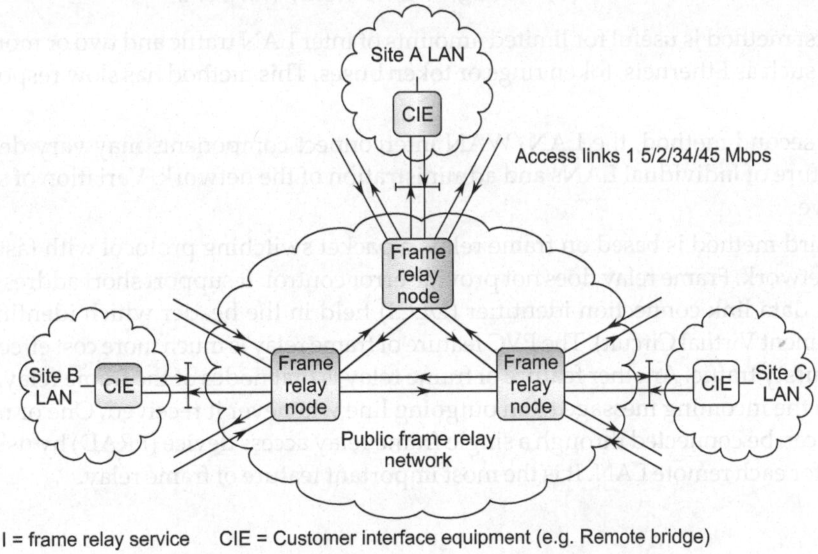

Fig. 6.8 *Public frame relay network*

Basically, there are three ways of LAN interconnection to implement Wide Area Network (WAN) *i.e.*, packet switching network through X.25, through Leased line and frame relay network, as shown in fig. (6.9).

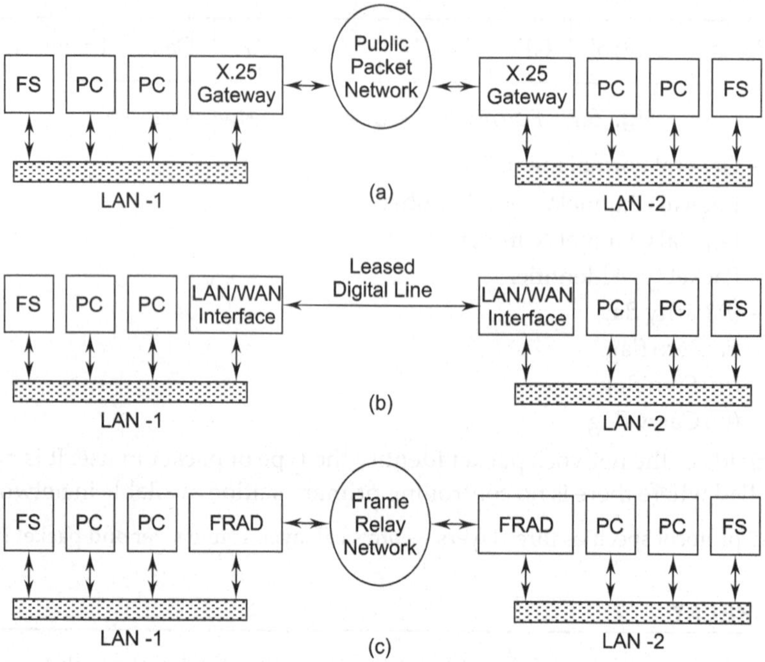

Fig. 6.9 *Three methods of LAN to LAN interconnection*
(a) Through X.25 Packet-switching network
(b) Through Leased Line
(c) Through frame Relay network.

- The first method is useful for limited amounts of inter LAN traffic and two or more dissimilar LANs such as Ethernets, token rings or token buses. This method has slow response and it is costly.
- In the second method, the LAN/WAN interconnect components may vary depending on the nature of individual LANs and administration of the network. Variation of speed is cost effective.
- The third method is based on frame relay, a packet switching protocol with faster rate than X.25 network. Frame relay does not provide error control. It support short address processing with a data link connection identifier (DLCI) field in the header which identifies the PVC (Permanent Virtual Circuit). The PVC feature of frame relay is much more cost effective for high rate, bursty traffic. Another feature of frame relay is that nodes of the frame relay can directly switch the incoming message to an outgoing line whenever it received. One or more remote LANs can be connected through a single frame relay access device (FRAD) by using different DLCI for each remote LAN. It is the most important feature of frame relay.

6.8 DIFFERENCE BETWEEN FRAME RELAY AND X.25 PACKET SWITCHING

Frame Relay	*X.25 Packet Switching*
(*i*) Link by link flow control and error control is not the responsibility of Frame Relay. (*ii*) Processing is done in two layers (layer 1 and 2) as switching is made by logical connection at layer 2 instead of layer 2. (*iii*) Call-controlling signalling is carried out on a logical connection separate from the data. As a result, intermediate nodes do not need to process call control messages on the basis of individual connections.	(*i*) *x*.25 packet switching handle the link by link flow control and error control. (*ii*) Three layers are required for processing as switching is made by logical connection that takes place at layer 3. (*iii*) Here, connection takes place on any one types of logical channels : (*a*) Switched Virtual Circuit (SVC) (*b*) Permanent Virtual Circuit (PVC).

6.9 TCP/IP

TCP stands for Transmission control Protocol, is the connection oriented protocol and IP or Internet Protocol, is the network protocol associated with internet.

TCP/IP was developed not only to create LAN'S but also for internet working of multiple LAN's. Different LAN'S with different protocols may be created and there different networks are connected through an internal gateway for interconnecting themselves. To connect with the Internet, external gateway is used. TCP/IP is the basic building blocks for the internet.

6.9.1 TCP/IP Reference Model

Reference model is simply a common framework for defining the specific set of protocols to be used with a particular application or network combination. The fig. (6.10) shows the TCP/IP reference model.

- TCP/IP is normally considered to be a four layer system. Layers of TCP/IP are: (*i*) Application layer (*ii*) Transport layer (*iii*) Internet layer and (*iv*) Network layer.
- The four layers stack of TCP/Ip deals with the equivalent seven layer architecture of OSI model.

Detail function of each layer is discussed below:

1. **Application layer:** This layer allow access to network resources through a suitable interface *i.e.*, it includes all processes and services that use transport layer to deliver data. Example include email, web access, telephony, video conferencing and so on.
   The most widely known application protocol are:
   TELNET, File Transfer Protocol (FTP), Simple Mail Transfer Protocol (SMTP) etc.
2. **Transport layer:** At the transport layer, TCP/IP has two protocols — Transmission Control Protocol (TCP) and User Data gram Protocol (UDP). Either one of the protocol is selected by application layer based on the services.

   In short transport layer enables to provide reliable end-to-end message delivery and error recovery. This layer also responsible for directing each information flow to and from the related application in case of multiple applications running at the same end system.

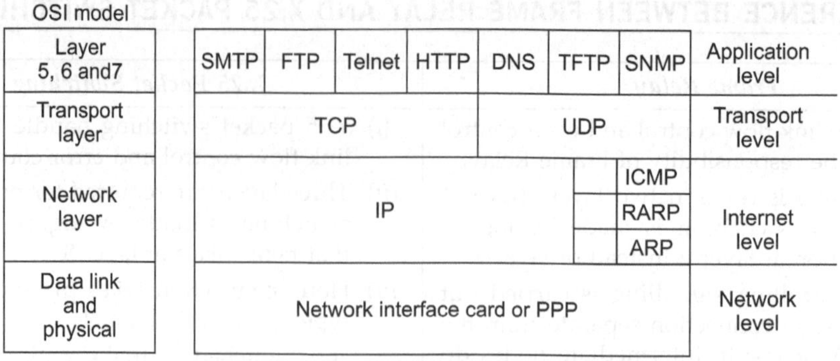

Fig. 6.10 *TCP/IP reference model*

3. **Internet layer:** This layer provide internet working *i.e.*, communicating between machines or end systems. Some used protocol of this layer are IP, ARP, ICMP (IP stands for Internet protocol, ARP stands for address Resolution protocol and ICMP named as Internet control Message Protocol). These protocols provide for transmission and reception of transport requests and handle network level control. The basic protocol/IP performs following services:
   (*i*) *Addressing:* Chose the proper route to deliver data to the destination end.
   (*ii*) *Fragmentation:* If there is large message which can't handled by the transport layer then this protocol breaks the messages into packets, each with unique internet address.

   This layer does not give any guarantee of the information, content in a packet but simply control how the packets is transferred.
4. **Network layer:** This layer is also referred as network interface layer. It is responsible for accepting and transmitting IP datagrams. This layer also used to provide mechanical and electrical specifications. Network layer is same as physical and data link layer of OSI model. It simply consist of a device driver in the operating system and the corresponding network interface card in the machine.

Protocols used for different layers:

1. **Application layer:**
   (*i*) SMTP (Simple Mail Transfer Protocol)
   (*ii*) Telnet
   (*iii*) FTP (File Transfer Protocol)
2. **Transport layer:**
   (*i*) UDP (User datagram protocol)
   (*ii*) TCP (Transmission Control Protocol)
3. **Internet layer:**
   (*i*) IP (Internet protocol)
   (*ii*) ICMP (Internet Control Message Protocol)
   (*iii*) IMAP (Internet Message Access protocol)
4. **Network layer:** No protocols are used for this layer.

6.9.2 Comparison between OSI and TCP/IP Model

OSI Model	*TCP/IP Model*
1. OSI model includes 7 layers.	1. TCP/IP model includes 4 layers.
2. This does not support internet working.	2. Support internet working.
3. OSI support connectionless and connection oriented communication in the network layer.	3. TCP/IP support only connection oriented communication in the transport layer.
4. Strict layering.	4. Lossely layered.
5. Reliable delivery of packets.	5. Unreliable delivery of packets.
6. Model was first defined and then implementation of protocol is done.	6. Protocol implementation was done before designing of model.
7. Not useful for today's advanced technology.	7. More useful for advanced technology.

6.9.3 Transport Level Protocols

Transport level includes two types of protocols: (*i*) User datagram protocol (UDP) and (*ii*) Transmission control protocol (TCP).

UDP: UDP provides a connectionless service as well as unreliable service also. It is an end-to-end transport level protocol that adds only port addresses (source address and destination address), checksum error control and length information to the data from the upper layer (*i.e.*, Application layer). The packet produced by the UDP is called a user datagram, shown in Fig. (6.11).

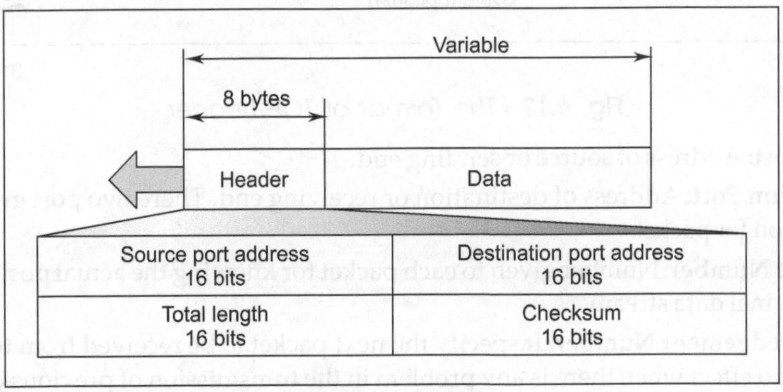

Fig. 6.11 *UDP datagram*

Source port address: The address of the application program that has created the message.

Destination Port Address: The address of the application programm that will receive the message.

length Information: This is the total length of the entire datagram in bytes.

Checksum: 16 bit field for detection of error.

TCP: Transmission control protocol provides a connection oriented, reliable, byte stream service. TCP collect the information packets from the application layer and establish a virtual circuit between sender and receiver until transmission ends. It assignee a sequence number of each packet. White transmission, TCP alert the each receiver first that datagram are on their way. When the receiving end gets the packet it sends acknowledgement to the sending end. If the sending end does not get the acknowledgement within a limited period it retransmit the previous packet.

The receiving TCP allows the other end to send as much data as the receiver has buffers for. The fig. shows the TCP data in the IP datagram and fig. (6.12) shows a brief description of TCP header format.

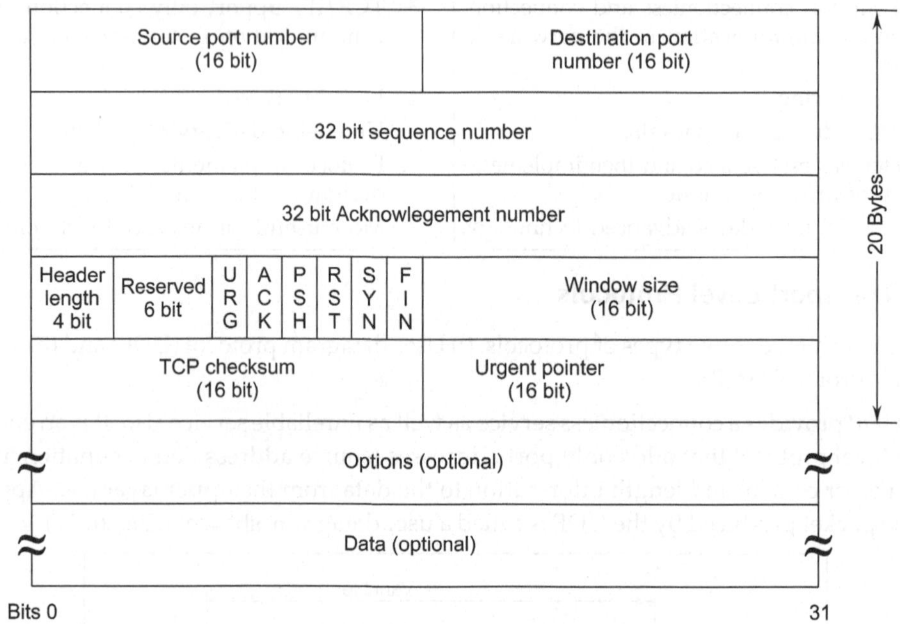

Fig. 6.12 *The format of TCP header*

1. **Source Port:** Address of source or sending end.
2. **Destination Port:** Address of destination or receiving end. There two port gives the routing information for packets.
3. **Sequence Number:** Number given to each packet for knowing the actual position of the data in the original data stream.
4. **Acknowledgement Number:** It specify the next packet to be received from the source. This field has no effect when there is any problem in the transmission of previous packet.
5. **Header length:** It is total length of the TCP header.
6. **Reserved:** This field kept vac for future use.
7. **TCP header includes six flag:** The function of each flag is as follows:
   (*i*) **URG:** The Urgent pointer is valid when set to 1 in case of urgent data to transfer.
   (*ii*) **ACK:** This flag set to 1 when acknowledgement number is valid.
   (*iii*) **PSH:** The PSH bit is used to inform the sender that data should pass as soon as possible.
   (*iv*) **RST:** Reset the connection to reject an invalid segment or at the end of a transmission session.
   (*v*) **SYN:** This flag synchronize the three types of segments—connection request, connection confirmation and confirmation acknowledgement with ACK flag.

   SYN = 1, ACK = 0 is a connection request.

   SYN = 1, ACK = 1 is a connection acknowledgement.

(*vi*) **FIN:** The FIN bit is used to release a connection or indicates that the incoming packet is the last packet.

8. **Window size:** Specify the number of bytes the sender is willing to accept.
9. **Checksum:** Used for transport layer's error detection.
10. **Urgent pointer:** If urgent flag set to 1, then this field, containing urgent data, define the end of urgent data and start of normal data.
11. **Options:** This is the optional field, used to provide other functions that are not covered by the header,

If URG = 0 = end of option list
1 = no operation
2 = maximum segment size.

12. **Data:** Contains user data, size is variable.

Fig. (6.13) shows the simple TCP operation

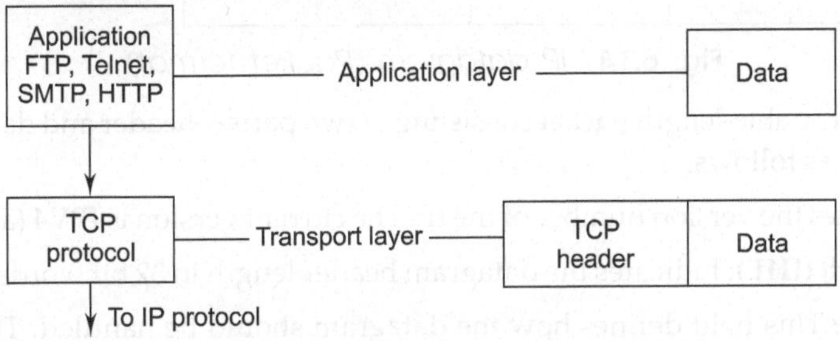

Fig. 6.13 *TCP operation*

6.9.4 Comparison between TCP and UDP

TCP	*UDP*
1. TCP is connection oriented.	1. UDP is connectionless.
2. TCP connection is byte stream.	2. UDP connection is message stream.
3. TCP is reliable.	3. UDP is unreliable.
4. TCP packet is known as segment.	4. UDP packet is known as user datagram.
5. TCP provides flow control and error control.	5. UDP does not provide flow control.
6. TCP does not support multicasting and broadcasting.	6. UDP support multicasting and broadcasting.
7. TCP supports full duplex transmission.	7. UDP does not support full duplex transmission.

6.9.5 Internet Protocol (IP)

The Internet Protocol or IP is a Internet layer protocol, containing address information and some control information. It is an unreliable and connection less datagram protocol—a best effort delivery service *i.e.*, it does its best to transmit the packets from source to destination but without any guarantee.

The IP of each end system has a unique Internet-wide address assigned to it. This is known as IP address. Each IP address has two parts—a network identifier (netid) and a host identifier (hostid).

The network identifier, identifies each network and the node or host address is assigned to and uniquely identifies each machine on a network.

IP transport data in packet forms known as datagram. Each datagram transported separately and may travel along different routes. IP is a connection service that's why it does not keep any record about the routing track and alerting call of delivered datagram. Fig. (6.14) shows the IP datagram.

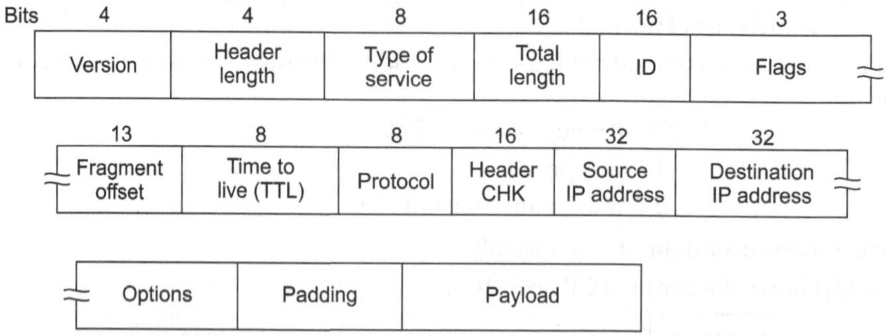

Fig. 6.14 *IP datagram (Packet format)*

A datagram is a variable-length packet consisting of two parts—header and data. A brief description of each field is as follows:

Version: It defines the version number of the IP. The current version is IPV4 (address size 32 bit).

IP Header length (IHL): Indicates the datagram header length in 32 bit words.

Type of service: This field defines how the datagram should be handled. The containing bits specify the type of service the sender desires such as reliability, delay or level of throughout and also defines the priority of the datagram.

Total length: Specify the total length of datagram (data and header). The maximum length is 65,535 bytes.

Identification: The identification field is used in fragmentation. A datagram may be divided into fragments to match the network frame size during transmission. Each fragment is identified with a sequence number in this field.

Flags: Contain 3 bits to deal with fragmentations DF = Do not fragment and MF = More fragment.

Fragmentation offset: It is a pointer that shows the offset of the data in the original datagram (if it is fragmented).

Time to live: Maintains a counter that gradually decrements down to zero, at which point the datagram is discarded. The maximum lifetime is 255 seconds.

Protocol: This field indicates which upper layer protocol data are given in the datagram (TCP, UDP, ICMP etc.)

Header checksum: 16 bit field verifies the header only. It is used to detect the error.

Source and destination address: 32 bit internet address specify the address of original source and destination.

Options: This is an optional field that control routing, timing, alignment and all other types of management.

Data: Contains upper layer information or user information.

Some Other protocols related to TCP/IP: The protocols listed here are used to support TCP/IP.

1. **ARP (Address Resolution Protocol):** Dynamically maps internet addresses to physical addresses on local area network (LAN).
2. **FTP (File Transfer Protocol):** Transfers files between hosts on the internet or TCP/IP network.
3. **ICMP (Internet Control Message Protocol):** A diagnostics and error-reporting and message control protocol between host and gateway at the IP layer.
4. **IGMP (Internet Group Management Protocol):** Manages multicosts groups in the internet.
5. **SNMP (Simple Network Management Protocol):** It used to gather information about devices on a network.
6. **RARP (Reverse Address Resolution Protocol):** RARP performs the reverse function of ARP. If there is host without disk then this protocol find the host's internet address at startup.

Note. Internet is composed of many networks (basically LANs) that communicate with each other using TCP/IP. So, that internet is another way for inter connecting LANs.

6.10 ATM (ASYNCHRONOUS TRANSFER MODE)

ATM is standardized architecture of packet oriented transmission and can support various serices like wide band, narrow band, broad band, bursty, real time, non real time etc. Basically, ATM is a telecommunication concept for the transport of broad range of user information. In fact, ATM can be thought of as the "highway" of the information superhighway. The data rate can be increased as in ATM, some software functions becomes hardware implementation.

Basically ATM is a cell switched network where the user access devices (known as end points) are connected through a user to network interface (UNI) to the switches. The switches are connected through network to network interfaces (NNIs). The architecture of ATM network is shown in Fig. (6.15).

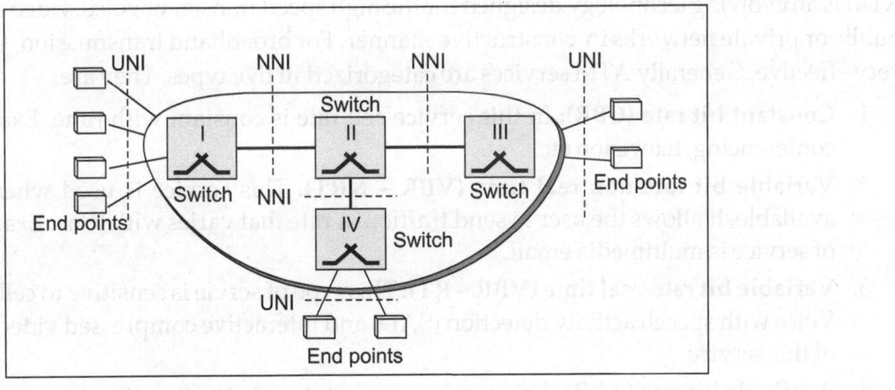

Fig. 6.15 *Architecture of an ATM network*

6.10.1 ATM Cell

Each ATM cell consits of 53 bytes in which there are 5 byte header and 48 byte payload. The format of each cell is shown in fig. (6.15). The header is made up of six fields. Their functions are as follows:

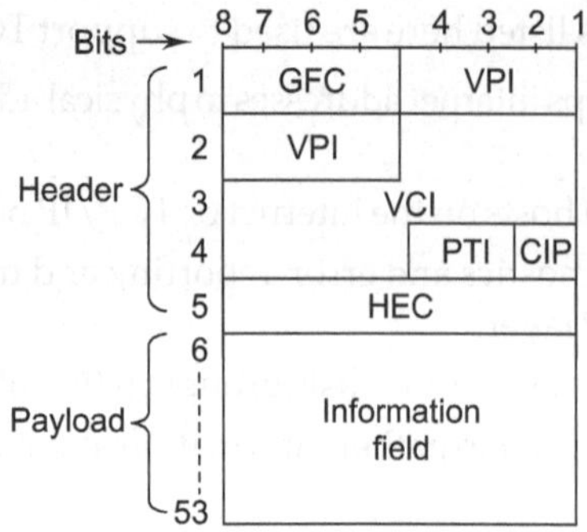

GFC = generic flow control

VPI = virtual path identifier

VCI = virtual channel identifier

PTI = payload type identifier.

CLP = cell loss priority.

HEC = header error checksum.

Fig. 6.16 *ATM cell formal (User network segment)*

- **Generic flow control (GFC):** This field confirms the entry of cell by a user into the network *i.e.*, it controls the flow of cell. It is preset only in the user-network interface (UNI).
- **Virtual path identifier (VPI):** It is used for identification or routing purposes within the network.
- **Virtual channel identifier (VCI):** It is also used for identification purpose. The difference between VCI and VPI is that VCI contains 16 bit whereas VPI contains 8 bits.
- **Payload type identifier/indicator (PTI):** This field indicates the type of information (whether text or control) contains the cell.
- **Cell loss priority (CLP):** During heavy load conditions, one or some cell (*s*) may be discarded. This field specicy that which cell should be discarded. If CLP = 0 then high priority (not to be discarded) and if CLP = 1 then low priority (should be discarded first).
- **Header error checksum (HEC):** This field check the errors of the header. It is a 8 bit field.

Note. A cell is a small data unit of fixed size.

6.10.2 ATM Service Categories

ATM is an evolving technology designed for the high speed transfer of voice, video and data through public or private networks in constructive manner. For broadband transmission, the use of ATM is very effective. Generally ATM services are categorized in five types. They are:

1. **Constant bit rate (CBR):** In this service cell rate is constant with time. Examples are video conferencing, television etc.
2. **Variable bit rate-non real time (VBR – NRT):** This service is used when information is available. It allows the user to send traffic at a rate that varies with time. Example of this type of service is multimedia email.
3. **Variable bit rate-real time (VBR – RT):** This type of servie is sensitive to cell delay variation. Voice with speech activity detection (SAD) and interactive compressed video are the example of this service.
4. **Available bit rate (ABR):** This service supports bursty traffic with a known minimum, mean and peak rate. It also provide additional rate to comp the demands on the avialable transmission capacity. If the network fail to provide additional rate then it informs the user to reduce the input rate. Examples are file transfer and email.
5. **Unspecified bit rate (UBR):** This service provide the similar service as provided by a best-effort network such as a LAN and the internet. It gives no guarantee to the user about the transfer of submitted information. This service is widely used for TCP/IP.

6.10.3 ATM Switching

Before sending any information, generally a virtual circuit is first established in the ATM network. A virtual circuit identifier used on each link is known as protocol connection identifier (PCI), subdivided into two fields—virtual path identifier (VPI) and a virtual channel identifier (VCI). Routing can be performed using either one or a combination of the two. So, basically two types of switches are there: (*i*) VP switch (*ii*) VPC switch. Based on there concept following types of switches are used:

1. ATM Memory Switch
2. Space Memory Switch
3. Memory space switch
4. Memory Space Memory Switch
5. Banyan Network Switch

Note. The VP switch route the cell using only the VPI and VPC switch route the cells using both VPI and VCI.

6.10.3.1 Memory Switch

The general structure of ATM switch is shown in fig. (6.17). There are input controllers (IC) which connects each input link separately through the input port. The input controller performs the routing of cells arriving at each input port to their corresponding output port by obtaining the correct output port number from the routing table. The port number is used to determine the path to be followed. Sometimes a situation arises where more than one link require the same output link. To sortout this problem cell buffers are used to hold the additinal cell (s) and follow the FIFO (First-In-First-Out) method to handle properly. The output controllers simply forward the received cell at the appropriate link.

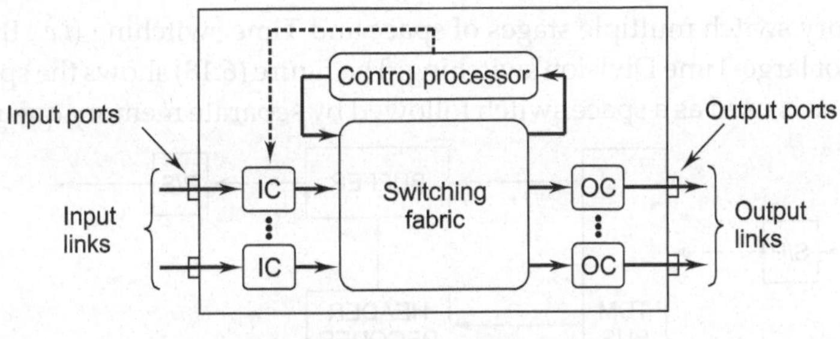

(a)

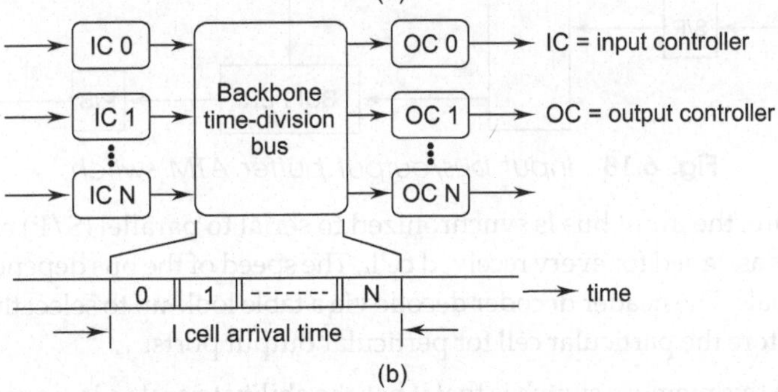

(b)

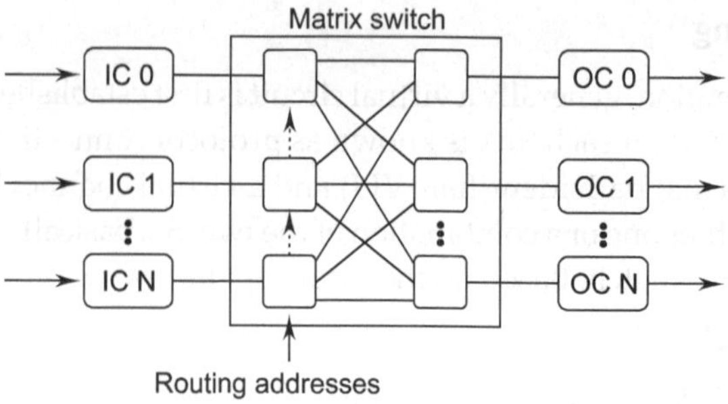

(c)

Fig. 6.17 *ATM switch architectures. (a) general structure (b) time division bus schematic (c) fully-connected matrix switch*

The main aim of the control processor is to download routing information into the routing tables in each input controller. The routing information is received through the network either from network management station or signalling control point processor if switched virtual circuit (SVC) is used.

Hence, after getting all the routing information, the cell is routed through the switching fabric directly from the input. In addition, the control processor performs other task like network management, fault reports, performs other task like network management, fault reports, performance statistics and so on.

A number of alternative switch fabrics are used in ATM switches. They can be classified as either time division or space division. It performs the synchronisation between the input and output.

Note. SVC or switched Virtual Circuit, is a connection that set up automatically by a signalling protocol. PVC or permanent Virtual Connection is a connection that set up manually by a network manager.

6.10.3.2 Space memory switch

In case of Space-Memory switch multiple stages of space and Time switching (*i.e.*, Buffering) are required as in the case of large-Time Division switching. The figure (6.18) shows the space-memory switch where a TDM bus is used as a space switch followed by separate memory queues.

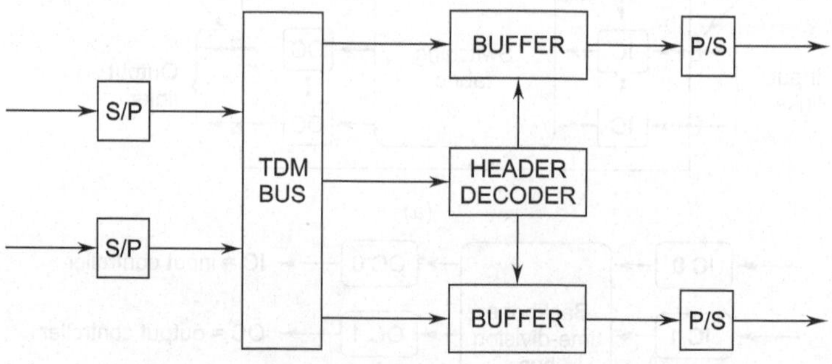

Fig. 6.18 *Input bus/output buffer ATM switch*

In this architecture, the input bus is synchronized to serial to parallel (S/P) circuits of the input ports as a time slot is assigned for every received cell., The speed of the bus depend on the total speed of the incoming signals. The header decoder decode via a table look-up to select the proper output so that the memories store the particular cell for perticular output ports.

The advantage of this memory switch is that it has the ability to scale a large range of switch sizes. But the main drawback is the memory speed.

To overcome the problem an advanced architecture is implemented where a high speed caches are used in between bus and buffer as a speed buffer. The advanced architecture is shown in Fig. (6.19).

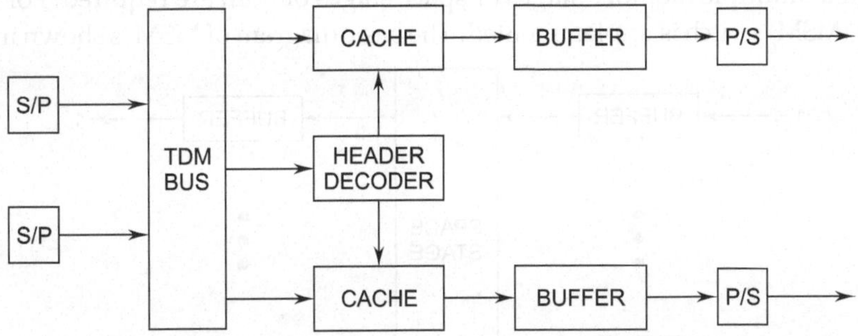

Fig. 6.19 *Input bus/output buffer ATM switch with catches*

6.10.3.3 Memory space switch

This switch is also known as Multiple Memory ATM switch as it use multiple stages is shown in figure (6.20). In this switch, all incoming cells are available on the output ports via separate paths for each input. There is a table look-up function attached to each buffer the select the appropriate cell for appropriate output port. The output port select the data from a separate buffer acts as a common queue or space switch function.

The number of buffer required is N^2 where N is the number of ports. The number of header processor is same *i.e.*, N^2.

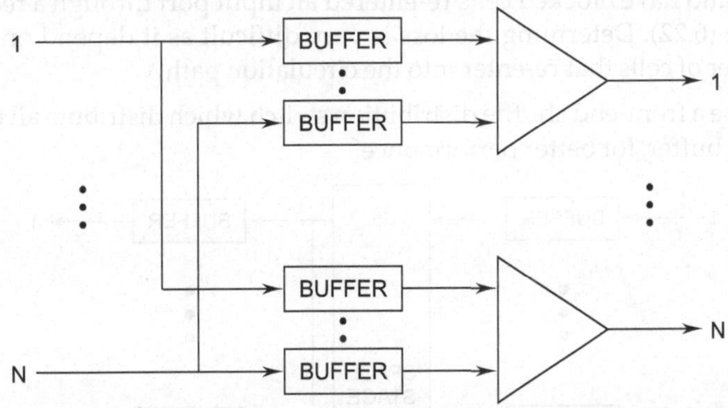

Fig. 6.20 *ATM switch with multiple memory*

If an incoming cell finds a full buffer then loss occurs. This situation can be controlled by encountering how much traffic of a particular output is accepted from a particular input.

The main advantage of Memory-Space switch is that all data paths and memories operate at the external link speed. As relatively low speed of memories cost is minimum. The only disadvantage is the large number of buffer (*i.e.*, N^2).

Note. For large ATM switches, Space-Memory switch and Memory-space switch are used but for low speed there switches are not practically used. Only in small system these switches can be used but possibiity of blocking is more.

6.10.3.4 Memory-space-memory switch

For a large system multiple memory stages or space stages or both are required. For that Memory Space Memory (MSM) switch is implemmented. The basic diagram of MSM is shown in figure (6.21).

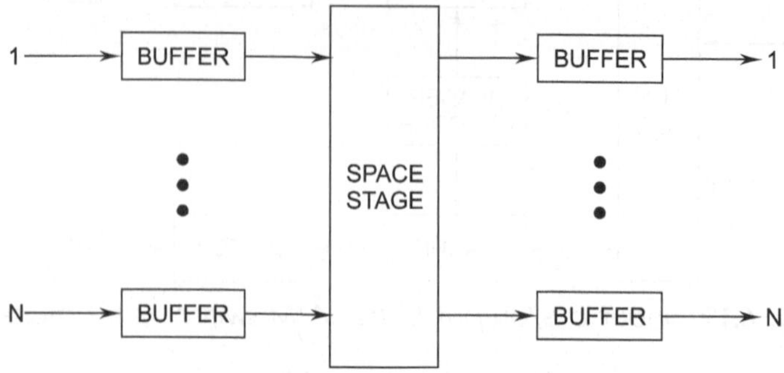

Fig. 6.21 *Memory-space-memory switch*

Like TST switch, the input memory stage collect the information that are queued for transfer through the space stage and then buffered again for transfer from the output memory. Before transfer a cell from input buffer to the output stage through the space stage, the output stage should be available for input. For proper transfer, a queue entry is made into an input buffer.

If the desired output is busy then blocking occurs. This type of blocking is known as head-of-line blocking. To minimize the blocking another cell in the input queue can be selected. Another approach is to determine whether there is a block in the output before transfer and then blindly transfer cells through the switch and have blocked cells re-entered an input port through a recirculation path as shwon in the figure (6.22). Determing the loss is very difficult as it depend on the probability of blocking and number of cells that re-enter into the circulation path.

So, there should be a front-end shuffle distribution switch which distribute all input lines of input cells across all input buffer, for better performance.

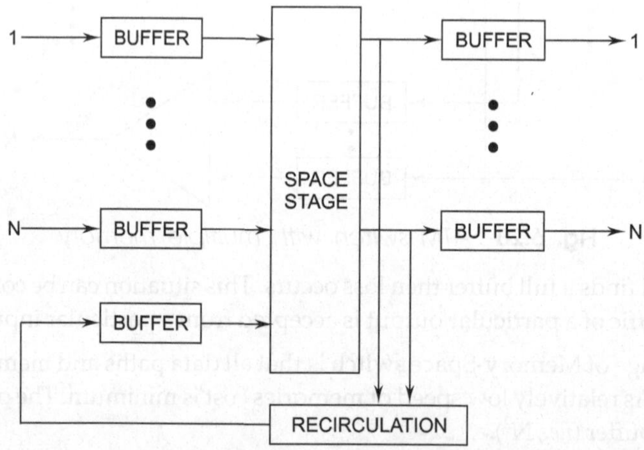

Fig. 6.22 *Memory-space-memory switch with recirculation*

To reduce the blocking probability, the architecture of MSM switch matrix can follow any one approaches as given as follows:

1. A design which includes a single-stage square matrix or
2. An architecture which includes a multistage class of service network.

Any design can reduce the blocking probability by running the space stage faster than the cell rate of the external ports. When multiple classes of service and multiple paths are accommodated in the switch then the design of hardware control logic become complex.

6.10.3.5 Banyan network switch

A more improved approach is a switch called a Banyan switch, named after the banyan tree. This is a multistage switch with microswitches at each stage that route the cells based on the output port represented as a binary string. If there are n inputs and n outputs then there will be log2 (n) stages with $n/2$ microswitches at each stage. Fig. 6.23. shows the banyan switch.

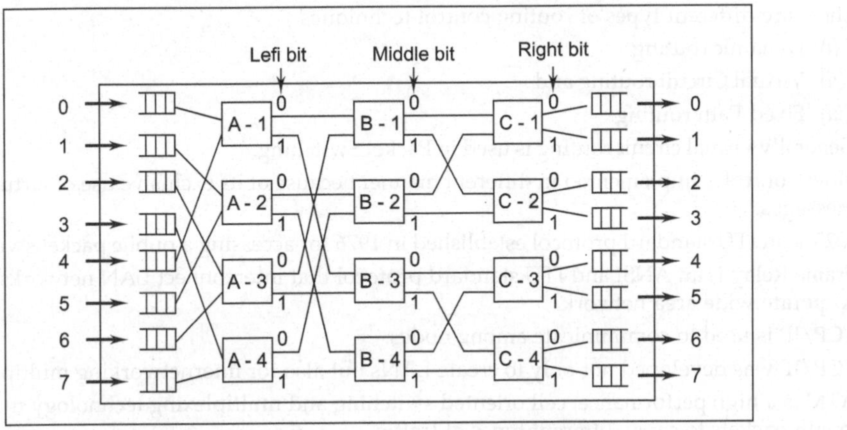

Fig. 6.23 *Structure of banyan switch*

The first stage routes the cell based on the high order bit of the binary string. The second stage then routes the cell according to the next second order bit and so on. The figure shows the eight inputs and eight outputs then the number of stages is log2 (8) = 3.

Operation: Suppose at the first stage a cell has arrived at input port 1 and has to go to output port 6 (110 in binary). The first microswitch A-2 routes the cell as the first bit is 1, second microswitch B-4 routes the cell for bit 1. Next the third bit is 0 and so, the third microswitch will be C-4. Thus, this is the routing procedure within the switch based on the binary string.

6.11 ADVANTAGES OF ATM

The advantages of ATM are as follows:

- A wide-range of information can be transmitted by the ATM network such s voice, video conferencing images etc.
- ATM can handle or control the wide range of information
- ATM is compatible with the optical fiber or twisted pair cable etc. as transmission media.
- ATM provde higher quality of service (QOS).
- The service is connection oriented as virtual circuit concept is used for data transfer.
- Any type of networking starting from generation to generation (LAN, WAN, MAN etc) can be integrated by ATM technology.

- Presently works with LAN and WAN technology by supporting protocols such as TCP/IP.
- Statistical multiplexing, a special type of multiplexing is used in ATM switching.
- The wide range of bandwidth (From 1.5 Mbps to 2 Gbps) is provided by ATM.
- The performance of the ATM network is high.

SUMMARY

- In Packet Switching, a single message is broken up into packets for transmission through the network.
- The first Packet Switching System was ARPANET.
- The main feature of packet switching operation is the manner in which the transmission links are shared on an as-needed basis.
- Statistical multiplexing is used in Packet switching.
- There are different types of routing control techniques
  - (*i*) Dynamic routing
  - (*ii*) Virtual Circuit routing and
  - (*iii*) Fixed Path routing.
- Generally virtual circuit routing is used in Packet Switching.
- Flow Control is implemented in different manner because of its exclusive use of virtual circuits for all messages.
- X.25 is an ITU standard protocol established in 1976 for accessing a public packet switching network.
- Frame Relay is an ANSI and ITU standard protocol that interconnect LAN networks to implement a corporate wide area network.
- TCP/IP is used to communicate among nodes.
- TCP/IP was developed not only to create LANs but also for internetworking multiple LANs.
- ATM is a high performance, cell oriented switching and multiplexing technology that utilizes fixed-length packets to carry different types of traffic.
- ATM switches provides two types of connection:
  - (*i*) Permanent Virtual Connection (PVC)
  - (*ii*) Switched Virtual Connection (SVC).

REVIEW QUESTIONS

1. Discuss the principle of packet switching.
2. Explain the packet format. Briefly discuss all the fields.
3. What do you mean by statistical multiplexing? Explain.
4. Compare dynamic routing with VCR and FPR.
5. What is X.25? Differentiate between X.25 and frame relay.
6. Draw and discuss the TCP/IP reference model.
7. What is ATM? Briefly explain different service categories related to ATM.
8. Explain different types of ATM memory switch.
9. Shortly explain different fields of ATM cell.
10. Compare between TCP and UDP.
11. Differentiate between OSI model and TCP/IP model.
12. Discuss the protocols related to TCP/IP model.
13. Discuss briefly about the routing control techniques.

14. Explain the terms:
(*i*) Header
(*ii*) Packet
(*iii*) Flow control.

MULTIPLE CHOICE QUESTIONS

1. Which one is not under the S and F message switching concept:
(*i*) Packet switching (*ii*) Circuit Switching
(*iii*) Message Switching (*iv*) None of the above.

2. Which type of routing has the possibility of greater transmission delay:
(*i*) Dynamic routing (*ii*) Virtual routing
(*iii*) Fixed Path routing (*iv*) None of the above

3. The primary requirement of ATM switching is to transfer cells arriving on one physical link to nother link:
(*i*) True (*ii*) False.

4. In Packet Switching which type of multiplexing is used:
(*i*) Synchronous Time Division Multiplexing (STDM)
(*ii*) Asynchronous Time Division Multiplexing (ATDM)
(*iii*) Frequency Division Multiplexing (FDM)
(*iv*) None of the above.

5. TCP provides flow control and error control:
(*i*) True (*ii*) False.

ANSWERS

1. (*ii*) 2. (*ii*) 3. (*i*) 4. (*ii*) 5. (*i*)

INDEX

A

Administration Software 89
Application Layer 126
Application Software 85, 89
ARP (Address Resolution Protocol) 132
Associated Signalling Mode 105
ATM (Asynchronous Transfer Mode) 132
Availability 76

B

Banyan Network Switch 138
Birth-death Processes 57
Blocking Probability 53
BORSCHT 46
Buffer 136
Busy Hour 50

C

Call Processing Software 90
Centralised SPC 77
Circuit Switching 4
Common Channel Signalling (CCS) 103
Common Control 75
Congestion 64
Consolidation 41
Crossbar Switch 9
Crossbar System 1
Cyclic Control 28

D

Datagram 119
Delay System 66
Digital Cross Connect System (DCS) 40
Digital Switching System 12
Direct Distance Dialling (DDD) 10
Distributed SPC 77, 81

E

Electronic Switching 11
End-to-end Signalling System 102
Ergodic Process 56
Ergostic Process 56
Erlang Traffic 62
Erlangs 51

F

Fill in Signal Unit (FISU) 109
Flow Control 123
Fragmentation 127
Frequency Division Multiplex (FDM) 97
FTP (File Transfer Protocol) 132

G

Grade of Service 53

H

Hard-wired Control 82
High Level Data Link Control (HDLC) Protocol 109
Holding Time 50
Horizontal Control 83

I

ICMP (Internet Control Message Protocol) 132
IGMP (Internet Group Management Protocol) 132
Implementation Complexity 33
Inband Signalling 97
Inchannel Signalling 97
Initial Address Message (IAM) 113
Inter-register Signalling 101
Interdigit Gap 94
Intermediate Distribution Frame (IDF) 8
Internet Layer 127

Internet Protocol (IP) 130

J

Jacobaeus Method 22

L

Lee Graphs 22
Line Signalling 101
Link Status Signal Unit (LSSU) 109
Link-to-link Signalling 101
Load Sharing Mode 78, 79
Long Distance Direct Current (LDDC) Signalling System 96
Lost Calls Cleared (LCC) 61
Lost Calls Held (LCH) 61, 65
Lost Calls Returned (LCR) 61, 65

M

Main Distribution Frame 8
Maintenance Software 90
Markov Processes 56
Mean Time Between Failures (MTBF) 76
Memory Based Control 28
Message Signal Unit (MSU) 111, 109
Message Switching 2
Microprogrammed Control 83
Multiple Stage Switching 16

N

Network Interface 124
Network Layer 127
Network to Network Interfaces (NNIs) 132
Network Traffic Load 49
Non-associated Signalling Mode 105

O

OSI Model 127
Outband Signalling 99

P

Packet Switching 3, 116
Part (TUP) 113
Poisson Arrival Process 60
Program Control Data Transfer 83
Program Status Word (PSW) 86
Pulse Dialling System 94

Q

Quasi-associated Signalling Mode 106

R

RARP (Reverse Address Resolution Protocol) 132
Register-sender 8
Reliability 76
Renewal Process 59
Rotary Dial Telephone 94
Routing Control 119

S

SDL (Specification Description Language) 87
Segregation 41
Signalling Information Field (SIF) 113
Single Stage Switching 16
SNMP (Simple Network Management Protocol) 132
Space Division Switching 16
Standby Mode 78
Statistical Multiplexing 118
Step-by-step or Stronger System 1
Stochastic Process 55
Stored Program Control (SPC) 1, 76
STS Switching 34
Subscriber Loop 93
Subscriber Network Interface (SNI) 124
Subscriber Traffic 51
Subscriber Trunk Dialling (STD) 10
Switched Virtual Circuit (SVC) 135
Synchronous Duplex Mode 78
System Software 85

T

TCP 126
TDM 100
TDM (Time Division Multiplexing) 4

Telephone User Part (TUP) 113
Time Assigned Speech Interpolation (TASI) 66
Time Division Multiplexing 100
Traffic Intensity 51
Transport Layer 126
Trunk 10
TSSST Switches 37
TST Switching 35

U

UDP 128
User to Network Interface (UNI) 132

V

Vertical Control 83
Virtual Circuit 119

X

X.25 Protocol 123

Z

Zero-loss Switching 43